Cast in Concrete

Concrete Construction in New Zealand 1850–1939

Cast in Concrete

Concrete Construction in New Zealand 1850–1939   *Geoffrey Thornton*

Also by Geoffrey G. Thornton:
New Zealand's Industrial Heritage (1982)
The New Zealand Heritage of Farm Buildings (1986)

The author and publisher gratefully acknowledge the assistance of the Stout Trust in the publication of *Cast in Concrete*.

Published 1996 by Reed Books, a division of Reed Publishing (NZ) Ltd, 39 Rawene Road, Birkenhead, Auckland 10. Associated companies, branches and representatives throughout the world.

This book is copyright. Except for the purpose of fair reviewing, no part of this publication may be reproduced or transmitted in any form or by any means, electronic or mechanical, including photocopying, recording, or any information storage and retrieval system, without permission in writing from the publisher. Infringers of copyright render themselves liable to prosecution.

ISBN 0 7900 0468 2

© Text and photographs copyright Geoffrey G. Thornton 1996

The author asserts his moral rights in the work.

Edited by Deirdre Parr.
Designed by Sally Hollis-McLeod, Moscow Design Ltd, Auckland.
Printed in Hong Kong by
South China Printing Company (1988) Limited

Front cover: Fairfield Bridge, Hamilton; Cape Foulwind Lighthouse.
Back cover: Cookshop, Castlerock station; water tower, Addington; Public Trust Office, Napier.

Contents

	Acknowledgements	6
	Introduction: A remote setting	8
1	The origins and development of cement and concrete	12
2	The early days in New Zealand	22
3	The 1870s	26
4	The 1880s	62
5	The origins of cement manufacture in New Zealand	86
6	The 1890s	90
7	The twentieth century: The first decade	98
8	1910–1919	128
9	1920–29	168
10	1930–39	210
	Conclusion: Utility and diversity	232
	Bibliography	234
	Index	237

Acknowledgements

Many people have assisted me in locating or providing material for the writing of this work. To them I express my gratitude. They are: Livingstone Baker, Patea; Alex Bowman, Nelson; Noel Crawford, Cave; Chris Cochran, Wellington; Bruce Collier, Whangarei; the late Rodney Draffin, Auckland; Alan Edwards, Dunedin; Jane and Rodney Foster, Kaitaia; Lois Galer, Dunedin; George Griffiths, Dunedin; the late Elaine Griffin, Wellington; Andrea Grieve, Christchurch; Michael Hitchings, Dunedin; Brian Jones, Perth, formerly Wellington; Ross Jenner, Auckland; Maurie H. Jerram, Taihape; Colin Kerr, Invercargill; Ian Lochhead, Christchurch; Ann McKewen, Christchurch; David McLeod, Christchurch; J.B. McAra, Waihi; Guy Natusch, Napier; the late H.C.N. Norris, Hamilton; David O'Kane, Port Hadlock, USA; Sheila Robinson, Gisborne; Beverley Simmons, Warkworth; Ian Stevenson, Opunake; Christopher Stanley, Slough, UK; Dorothy Tanner, Timaru; Pam Wilson, Christchurch; Kitty Woods, Wellington.

The following organisations have kindly provided data and in some instances drawings for which I am most grateful. I have used the original official name at time of writing: Auckland City Council (R.D. Gross); Auckland Harbour Board (B.R. Le Clerc, Chief Engineer); Eltham District Council (D.R. Bevan, District Manager); Hamilton City Council (Water Division, C. Atutahi); Hawera District Council (R. Moody, Acting District Engineer); Hurunui County Council (B.M. Yates, County Engineer); Inglewood County Council (L.J. Barnes, County Engineer); Manukau City Council (J. Weir, City Engineer); Ministry of Works & Development, Tauranga Residency (Mrs E.R. Wood); New Zealand Cooperative Dairy Coy Ltd (I.M. Calvert, Group Manager); New Zealand Electricity Division, Ministry of Energy; Rangiora Anglican Parish (Elizabeth Smith); Rangitikei County Council (P.S. Smart, Asst Planner); Southland Museum (Russell Beck, Director); Stratford County Council (M. Oien, Engineering Officer); Strathallan County Council (L.J.S. Baker, County Manager); Taranaki City Council (R.W. Struthers, City Engineer); Taranaki Museum (Ron Lambert, Director); Te Aroha Borough Council (R.M. Rankine, Borough Engineer); Thames & Coromandel District Council (Ross Vincent, Chief Engineer); Tuapeka County Council (B.F. Coleman, County Clerk); Vincent County Council (G.M. Smith, County Clerk); Waikato Art Museum (Rose Young, Curator, History); Wanganui City Council (P.M. Barnes, City Engineer); Wellington City Council (R.E. Lane, Design Engineer); Winstone Concrete Ltd (B.J. Ensor, Marketing Manager); Works & Development Services Corporation NZ Ltd (Evan Burt).

To all those who allowed me to photograph their buildings I express my thanks. They are: Mrs Jules Adams (Water Lea, Auckland); Mr and Mrs J.K. Carter (4 Kingsview Road, Auckland); Chris and Lesley Duncan (Overton, Marton); R.C. Gordon (Taihape Freezing Works); Brian Hawke (White Rock Woolshed); Cliff Matchett ('Bowmar House' Mangawhai); Mrs Diana McDonald (Whare Ra, Havelock North); Donald and Noeline McLaren (Mercer Road, Puerua); Mr and Mrs J. Roy (Maze

House, Pleasant Point); Mrs Phyllis Spencer (74 Te Nga Wai Road, Pleasant Point); Mrs Carol Scully (1 Grosvenor Terrace, Wellington); Rob Shand (Strathconan, Fairlie); Mrs Beverley Simmons (Riverina, Warkworth); Ian Stevenson (Pettigrew House, Opunake); Evan Upritchard (Ardross, Amberley); Richard Warnock (350 Richmond Road, Auckland).

I thank the staff of the various libraries consulted: Alexander Turnbull (National Library, Wellington); Victoria University of Wellington; Wellington Public (Central); University of Otago (Hocken); and the Cement & Concrete Association, Slough, UK.

In particular I am very grateful to Gavin McLean of the New Zealand Historic Places Trust; Professor Robert Park of the School of Engineering, University of Canterbury; Professor Peter Lowe of the School of Engineering, University of Auckland; my colleagues of the National Committee for Engineering Heritage, IPENZ, especially John Pollard, George Mullenger and the late Malcolm Jones for their strong interest and support.

Finally, I acknowledge the enthusiastic acceptance by Reed Publishing (NZ) Ltd of this work, and in particular Ian Watt as Managing Editor and Deirdre Parr as Editor.

It should be noted that the private properties illustrated are by courtesy of the owners and are not open to the general public.

G.G.T.

Author's note

Over the period covered by this book, the names of individual companies often varied considerably — both within their own records and as a result of company mergers and changes in structure or ownership. Whenever possible I have referred to the official company name for the period under discussion. Where more than one version existed, the name has been standardised for consistency.

Introduction
A remote setting

The term 'concrete' has come to have several meanings in the English language. It is a fourteenth-century derivation from the Latin concretus meaning 'grown together, hardened'. As a building material it has a very long history, although it did not come into general use until late in the nineteenth century. It is remarkable that New Zealand, as remote a country as any in the Victorian period, should have made very considerable use of concrete in all manner of structures. What is also surprising is that much of the early building was by individuals rather than by government and local bodies.

Although there are no recorded cogent reasons for this use of an innovative building material in a raw young colony with a small population, the most likely explanation is the shortage of skilled tradesmen such as stonemasons and bricklayers in the early days. Timber was widely used, being the predominant material for houses, much industrial and commercial work, and at first even for public buildings. In the country districts farm buildings were mostly of timber except in Central Otago and South Canterbury. Timber, however, was often associated with temporary structures, whereas concrete suggested permanence.

Another possibility is that the pioneering instinct resulted in a greater readiness to experiment with new ideas rather than simply accept traditional attitudes to building materials. It should be remembered that many immigrants were skilled people. They had been stifled from fulfilling their potential for expression of new technological developments in the overcrowded conditions prevailing in Britain.

Perhaps it was the fear of earthquakes and fire which made a strong case for the use of concrete. Certainly fire was a very real threat to nineteenth-century timber structures, many of which have burned down over the years. There were also some do-it-yourself experiments with concrete where the advice and experience of trained architects, engineers and builders was absent.

For many people concrete brings to mind a heavy, dull grey, uninteresting material having great strength but no beauty. But what exactly is concrete? The fundamental component is cement, which is a finely ground, grey mixture of calcined limestone and clay. Cement unites a mix of relatively small units of inert material, of which the coarser fraction may be broken stone, brick, shingle or gravel and is known as coarse aggregate. To fill the smaller interstices a finer material is added, usually sand, referred to as the fine aggregate. After dry mixing of these ingredients sufficient water is added to start the chemical action of the cement and to make a suitably plastic mixture. When thoroughly prepared, originally by hand mixing but by mechanical means today, the concrete is placed in prepared forms known as formwork, boxing or shuttering, constructed of timber or sometimes steel, which support or contain whatever element is being made — floors, walls, roofs,

SCAB DIP, LANGLEY DALE STATION, MARLBOROUGH.
The dip was used in the 1870s to treat sheep against scab disease. A gate with a tipping device helped the sheep to slide into the pot-type dip.

beams, columns, arches and so on. Although concrete hardens slowly, the formwork may be removed after a few days depending on the particular circumstances. For beams and slabs a rather longer period is necessary than for walls.

For many years plain concrete was the only form known. This has the property of great strength in compression — that is, the ability to support vertical loads. Conversely it has little resistance to tensile forces such as bending. To overcome the problem some early experiments in the mid-1850s introduced iron into the concrete. The result was that walls could be made thinner and stronger, floors could be suspended and the system of column and beam construction could evolve. Nevertheless it was a long time before the idea of reinforced concrete became a practical reality and in the nineteenth century many structures were built of plain concrete. Understandably engineers limited its use to foundations, docks, bridge abutments, piers and suchlike where it was sometimes referred to as mass concrete. Although England was the home of plain concrete and the originator of Portland cement production, comparatively little evidence remains there of early examples in this material.

Concrete surfaces may be left in 'off-the-form' state (as stripped from the boxing) or rendered with a cement plaster. Sometimes lime was used for plaster. For more presentable appearances a veneer of stone or brick was added as an exterior finish, especially on important public and communal buildings.

The effects of weathering can be avoided, or at least minimised by frequent washing with water, which assists in keeping the surface relatively free from staining. However, the particular texture of the concrete surface affects the effectiveness of this treatment.

In New Zealand, early concrete structures usually relied on their inherent colour and texture. Over the years the concrete has darkened very considerably, sometimes to a charcoal grey, giving the buildings a remarkable uniformity of colour. While some people regard this as forbidding, others find it quite attractive.

MAGAZINE, LYTTELTON HARBOUR.
The magazine was built in 1874 at what became known as Magazine Bay, and was used initially to store explosives for non-military purposes such as rock blasting. The roof is brick. It is unusual to see buttresses in such a small concrete building.

As early as 1871, in a debate on concrete at the Royal Institute of British Architects in London, Sir Arthur Blomfield advocated off-the-form surfaces without the need for plastering. He stated that construction joints and formwork marks should be used to give colour and texture to the concrete surface. His ideas were advanced for the time and went unheeded for the next 60 years.

From the 1930s attention has been given to the finishes to concrete surfaces. Careful planning of joint lines to produce a pattern, cleaning down the surface by grinding protuberances and filling voids, using exposed selected aggregates to give colour and/or texture, nailing strips of wood to the formwork to make patterns in the concrete, and using textured liners are some of the methods that have been employed. In New Zealand the effectiveness of such treatments can be seen in hydro-electric power stations and in motorway bridges built in the 1950s and sixties.

The reinforcement of concrete is achieved by embedding steel rods or bars so that they bond with the concrete to act cohesively under stress. In beams, the steel is placed at the lower level as this is the position most effective in providing resistance to bending. There is a tendency for beams to shear or crack diagonally near their supports. To overcome this the steel is bent to cross the shear line in a cranked form.

Steel is necessary in concrete columns to resist the tendency to buckle under load. Vertical rods are used and tied together at intervals by stirrups wired to them. Slabs for floors and roofs, and also walls, have reinforcement — usually a two-directional mesh of rods. Floor slabs on the ground require a minimum of steel to help distribute the shrinkage stresses and today spot-welded mesh of small gauge is often used.

The technology of concrete has become highly complex over the years and it is not the purpose of this book to go into

APRON AND STEPS, WANGANUI RACING CLUB STAND.
This structure was built in 1876 and has been modified in more recent times. The broad sweep of the shallow steps is in keeping with the original design.

BRIDGE IN GARDEN AT ELDERSLIE, NORTH OTAGO.
The grounds of this former grand estate were laid out to a plan by Sir Joseph Paxton, designer of the Crystal Palace, and the bridge may be his work. It was probably built in the 1870s.

the technicalities. Rather the intention is to introduce the reader to types of structures that were built in this country from the earliest use of concrete up to the outbreak of the Second World War in 1939.

Almost all of the examples given and illustrated are still standing with most still being used. As far as I have been able to ascertain, the development and use of plain concrete in New Zealand was rather more widespread than in most other countries in the nineteenth century. It is remarkable that plain concrete was widely used for farm buildings, many of which were remote from towns. There are also pioneer efforts dating from the 1870s in elementary reinforced concrete. This is a record of which any country can be proud. Reinforced concrete, at first often referred to as ferro-concrete, was a new development that New Zealand was quick to adopt, resulting in some very worthy structures being built in the early twentieth century.

The material in this survey is in chronological order, rather than by types of structure, so that the broad historical development of concrete construction in this country can be more effectively presented. Because of the wealth of extant concrete structures from the turn of the century I have had to be selective. At the time of writing, however, some of these examples are being demolished with little or no thought given to their historic and technical importance.

Chapter One
The origins and development of cement and concrete

For many years it has been generally accepted that the discovery of cement was of Roman origin. Its particular properties had come to the notice of the inhabitants of ancient Putoli (Puzzoli) who used the local volcanic sand with their lime mortar. This pozzolanic cement, as it became known, had the capacity to make the mix set harder and more rapidly. Furthermore it was able to set under water. The harbour at Putoli, built in 199 BC, made use of this property and from that time concrete became the main structural building material of the Romans, although it was commonly covered in a skin of either stone or brick.

In early 1986 it was reported from Beijing that archaeologists in northwest China had discovered a large concrete floor predating the Roman use of cement. Said to be 5,000 years old (c. 3000 BC) from the carbon dating technique, it goes back to neolithic times. The report states that this concrete floor in Dadiwan, Qinan County, Ganow Province, has similar ingredients to present-day concrete as it used sand, stone, broken pottery, and bones giving silica and aluminium.

In his booklet *Highlights in the History of Concrete* (1980) Christopher Stanley states that the oldest concrete so far discovered goes back to c. 5600 BC. Some excavations at Lepenski Vir on the banks of the Danube in Yugoslavia revealed a 250-mm-thick floor composed of a mixture of red lime, gravel and sand to which water had been added. It is believed to be the floor of a hut that had a framework of close-spaced wooden poles for cladding and roof. Stanley also mentions that the earliest pictorial evidence of the use of concrete appears in a mural dating from c. 1950 BC at Thebes in Egypt showing various stages in the making of concrete and mortar.

Ancient Roman concrete was usually prepared by pouring pozzolanic cement on layers of small broken stones (*caementa*) and repeating the process until the structure was filled. Baths and other buildings had used concrete vaults from the first century BC. The Pantheon in Rome (built AD 118–28) with its splendid 43-m-diameter dome portrays the grandeur of Roman architecture and its structural concepts. The dome is built of lightweight concrete using crushed pumice aggregate. Concrete was widely used right to the end of the Roman Empire.

The Roman writer Vitruvius, in his books on architecture c. AD 27, gave lengthy descriptions of the properties of concrete as well as the various forms of pozzolanic earths. In Roman Britain local materials were used to produce what is regarded as a lime concrete. Examples still exist today, such as Hadrian's Wall in the north of England (AD 122–30), where the core of the stone wall was bonded with concrete.

After the fall of the Roman Empire the knowledge of concrete making and construction was lost. During the Dark Ages (late fifth century to AD 1000) it was limited to a few isolated instances with the skills of manufacture being passed from father to son. It is believed that concrete was reintroduced to Britain by the Normans, who used it in foundations for castles and ecclesiastical buildings.

During the Renaissance Fra Gioconda (c. 1433–1515) edited a work on Vitruvius. In his construction of the Pont de Nôtre Dame in Paris in 1499 he made use of pozzolanic mortar in the piers, claiming to be the first on record to have used concrete since Roman times. However, a pozzolanic material known as Rhenish trass was used for many of the protective works in the Low Countries.

Development in England

It was not until the late eighteenth century that an interest in concrete was revived. In 1759 John Smeaton, a Leeds engineer, was given the task of replacing the Eddystone Lighthouse — a timber structure that had blown away in a gale. He used blocks of stone bonded together with a cement mortar he had produced after experiments for hardening under water. His successful cement was a mix of Welsh blue lias (limestone) and an Italian pozzolana which was also used as the base of the lighthouse. Smeaton wrote a book on his researches into cement entitled *A Narrative of the Eddystone Lighthouse*.

Various attempts to develop new forms of cement followed Smeaton's work. A clergyman, Rev. James Parker of Northfleet, Kent, discovered by chance that the Thames Estuary beach stones, which were nodules of septuaria, calcined under heat and when crushed to a powder produced a cement. He took out a patent in 1796 calling his product 'Roman cement' as he believed it to be the same as that used by the Romans. The first recorded example of its use was in the London warehouse of the British Plate Glass Coy.

In 1813 a copy of John Smeaton's book was bought by a young Leeds bricklayer, Joseph Aspdin. This must have stimulated his interest in cement as did a patent taken out in 1811 by Edward Dobbs. On 21 October 1824 Aspdin patented the first Portland cement, so named because its colour was similar to that of Portland stone. His patent was entitled 'An Improvement in the Modes of Producing an Artificial Stone', the cement being produced by burning a controlled mixture of clay and limestone. It was undoubtedly a superior cement but intended basically as a plaster rendering over brickwork to imitate Portland stone.

Joseph Aspdin had done his research in Wakefield and it was there that he set up his cement works at Kirkgate, having as a partner William Beverley, a brassfounder and tinplate manufacturer. In 1843 Aspdin's elder son, James, entered into partnership and in the following year took over the business and continued the manufacture of Portland cement for the next eighteen years, until his death. However, the firm stayed in business as Aspdin & Coy until 1904. It is of interest to note that as early as the 1830s an American, Obadiah Parker of New York, had produced a cement similar to Aspdin's and he built several houses of this material with monolithic walls.

In 1840 Charles Francis and Sons produced Medina cement, which they described as an improved Roman cement, being three times stronger. Their works were established at Parkhurst on the Isle of Wight and a remnant can be seen today in the small office building. There are some old kilns too, but these are most likely later structures. In 1852 one Langley expressed his faith in this material by building two houses at East Cowes on the isle. Four years later Charles Francis and Sons built two concrete houses with their own cement for the Army authorities, who expressed satisfaction at their cheapness, rapid erection and comfort as quarters. Apparently earlier attempts to use Roman cement for small buildings in London proved unsuccessful because the cement was too low in strength.

Joseph Aspdin had another son, William, who established his own cement manufacturing business in the lower Medway Valley at Rotherhithe. The renowned engineer I.K. Brunel used Portland cement supplied by William Aspdin to seal damage to the roof of the Thames Tunnel and also in its partial relining. In 1847 Aspdin set up a new cement works at Northfleet in partnership with two businessmen. Five years later he moved to Gateshead-on-Tyne to build a cement works which was the largest in the world at the time. In 1860 he left this venture to begin manufacturing cement in Germany where he died four years later.

In 1837 John Bazley White and Sons set up a cement manufacturing business with offices in London. J.B. White had been a partner of Charles Francis in 1808 in the manufacture of Roman cement. In 1838 the Whites appointed as their works manager the young Isaac Charles Johnson, who remained with them for fourteen years. John Bazley White has the distinction of having built the very first all-concrete house, at Swanscombe, in Kent, in 1835. Walls, tiles, window frames and decorative work were all in concrete. There were also concrete gnomes in the front garden. Another early use of concrete was in the substantial country house built for Queen Victoria on the Isle of Wight between 1845–48.

When I.C. Johnson left the Whites in 1852 he established his own cement works near Rochester. Four years later he moved to Gateshead, taking over the cement works of William Aspdin. Here the firm of I.C. Johnson and Company made

Portland cement, Roman cement, Keene's cement and plaster of Paris. Johnson improved on Aspdin's product with his 1844 patent wherein he heated materials of closely controlled chemical composition to a sinter temperature of 1,398°C. The resultant more reliable cement was not dissimilar to that of today.

After Johnson left J.B. White & Sons in 1852 that firm became J.B. White & Brothers, although John Bazley retired at that time. Johnson's successor was William Goreham. A joint patent with Leadham White taken out in 1870 was the most significant improvement in the manufacture of Portland cement since its invention. This consisted of mixing chalk and clay with only one-fifth to one-third of their weight in water. The much thicker slurry produced was passed through millstones to disintegrate the coarse particles (the process being known as wet grinding) and was then pumped direct to kiln drying chambers. This method obviated the need for settling tanks and so shortened the manufacturing time by several weeks, as well as reducing the area required for a cement works.

Two years later (1872) I.C. Johnson patented his chamber kiln, which was like a bottle kiln but used a horizontal arched chamber on piers up to 30 m in length. Slurry was pumped direct from the wash mill to the top of the kiln chamber to fall through inlets to the chamber floor. When it was about 200 mm thick, the inlets and flues were sealed, the kiln was fired, and gases passing along the chamber and out of the chimney gradually dried out the slurry. Johnson later modified this to enable gases to pass under the slurry as well. As the most successful kiln of its type it became very popular.

For many years England was virtually the only country producing Portland cement. The lack of overseas competition caused public indifference to its use and an unwillingness to develop knowledge of the chemistry of cement. However, this started to change when an engineer, John Grant, who was using Portland cement in sewer construction for the Metropolitan Board of Works in London in 1858, made many tests for compressive and tensile properties using samples from various manufacturers in the form of moulded briquettes. The results were published in three papers presented to the Institution of Civil Engineers in 1865, 1871 and 1880.

In 1862 England produced 203,200 tonnes of Portland cement. The biggest manufacturer, Whites, contributed 30,480 tonnes and employed 750 men, with most of their product going to France and Germany. From this time cement was also exported to India, Australia, New Zealand, South Africa, Canada, the USA, South America and Russia. In 1854 Germany had three cement works but by 1882 there were 420 and by 1895 this number had increased to 1,274.

In 1856 England, France and Germany were using Portland cement in docks, harbour basins, fortifications and for the elevations of houses in imitation of stone. Another early use was for coating rivets on the insides of iron ships to prevent corrosion by bilge water. The earliest use of plain concrete in Britain was in 1800 when William Jessop used Parker's Roman cement in the construction of the West India Dock. Portland cement does not appear to have been used for buildings much before the 1860s. One reason could be the prohibition on this new and unproven material by the Metropolitan Board of Works.

The United States imported cement from Britain until 1871. In that year David O. Saylor started to manufacture artificial Portland cement at Coplay in Pennsylvania. His company had been established in 1866 and the cement works ceased about the turn of the century.

Reinforced concrete

The origins of reinforcement in concrete are a little contradictory in some respects. The mix of sand, stone and cement with water produces a concrete which is weak in tension and so early attempts were made to find a satisfactory way of overcoming this drawback. Various materials were embedded in it where structural stability was required. A British engineer, Ralph Dodd, took out a patent as early as 1818 for the inclusion of wrought iron bars in concrete (at that time made with Roman cement). The great British engineer Thomas Telford (1757–1834) used iron bars in the concrete abutment of the Menai Straits Bridge, built in 1825. In 1829 a Dr Fox with one Barrett had a method, patented in 1844, of filling the space between cast iron I-beams with concrete. During the 1850s there were many patented inventions for combining iron with concrete.

One of William Aspdin's customers, William Boutland Wilkinson of Newcastle, England, had been a plasterer and manufacturer of artificial paving stones. He must have been an intelligent man for he took out patent rights in 1854 for 'Improvement in the construction of fire-proof dwellings, warehouses and other buildings and parts of the same'.

His specification showed a remarkable understanding of concrete construction and theory by his method of erecting floors and ceilings with a network of flat iron rods (hoop iron) or second-hand wire ropes embedded in a concrete which was to include Portland or other cement of equal quality. This patent suggests that Wilkinson possibly had two firsts to his credit — he may have been the first person to specify Portland cement concrete in structural work, and he was undoubtedly the first to introduce the real principles of reinforced concrete construction. Surprisingly his achievement had little or no effect on the building industry for more than 30 years. Wilkinson was a pioneer ahead of his time. Unfortunately the only known evidence of his reinforced concrete work was a small cottage built in 1865 and demolished in 1954. It was noted then that the reinforcement was in the correct position within the beams and slabs so as to resist efficiently the tensile forces that occur in bending. The first-floor slab was formed of precast plaster moulds supported in position, between which colliery ropes were laid in a regular pattern. Spaces between the moulds were filled with plaster and covered by a concrete slab with a granolithic top. There were beams 660 mm apart and 165 mm in depth. These were reinforced with 10-mm twisted wire ropes.

In 1849 Lambot, a Frenchman, built a boat of concrete reinforced with iron rods and exhibited it at the 1854 Paris Exhibition. This craft was in use well into the twentieth century although the techniques of its construction were not followed up for many years. In 1855 there was a patent (the first of many) issued to François Coignet for a system of combining iron joists with concrete.

After Wilkinson the next serious attempt to use reinforcement in England was by Joseph Tall in 1866. He built two similar cottages at Bexleyheath in Kent using a patented process for casting monolithic walls of concrete. The cottages, still standing today, had floors with a mesh of hoop iron embedded in the concrete slab.

A Parisian gardener, Joseph Monier (1823–1906), first used iron mesh as reinforcement in 1861 for his concrete flower tubs. He patented this idea in 1867 and used it in containers, pipes and railway sleepers. In 1875 he built a 16-m arched bridge (possibly the first use of reinforcement in a bridge), but his floor slabs were less successful due to inefficient positioning of the reinforcement.

In the United States a British lawyer, Thaddeus Hyatt (1816–1901), experimented independently with flat iron bars perforated at intervals to receive transverse round bars. He deduced that such an arrangement should be placed in the tension zone of concrete beams or slabs. Hyatt advocated T-beams and also made remarkably early deductions on the behaviour of reinforced concrete. He returned to England, undertook further testing in a laboratory in 1876–77, and then published a pioneer work, *Experiments with Portland cement concrete*.

Another Frenchman, Anatole de Baudot (1834–1915), designed and built the church of St Jean de Montmartre in Paris in 1894, using slender concrete columns and vaults with thin walls enclosing the structure.

However, it was François Hennebique (1842–1921) who is acknowledged today as the great French pioneer in reinforced concrete. Having begun his study in 1879 he soon established that the proper position for reinforcement in a concrete section such as a beam was in the lower portion. Possibly he knew of Hyatt's work. Hennebique was the first person to use steel, rather than iron, as reinforcement. His ideas on reinforced concrete were patented in 1892 in France and Belgium after his first building in this material the previous year. These patents came to be used in Western Europe and the United States. For several years he was the contractor for the erection of his reinforced concrete buildings, and then became a consultant engineer. He reasoned that to encourage confidence in the new material, the design and calculations should be done on a strictly professional basis and be remunerated independently of the contractor's profit and, furthermore, that the primary conditions for successful reinforced concrete construction must be impeccable workmanship and constant supervision. To achieve these aims he affiliated his organisation to several reputable building contractors in whom he had complete trust, ceding them concessions to use his patents under strict conditions.

In 1895, to get his system established in Britain, Hennebique sent the engineer L.G. Mouchel to undertake the contract for Weaver's Granary and Flour Mill in Swansea. This was the beginning of a prodigious amount of reinforced concrete construction in England using his patented systems. In all, 36,809 structures including buildings, bridges, viaducts, maritime structures, reservoirs, water towers, and canal works were completed between 1897 and 1919. One of these structures, built in 1900, was the first water tower of reinforced concrete in Britain, being that at Meyrick Park in Bournemouth and still in existence. Hennebique's European bridges included the Pont Neuf Chatellerault (1898) spanning 50 m and the Risorgimento Bridge (1911) in Rome of 100 m span.

An English expatriate in the United States was Ernest Leslie Ransome (1844–1917) who had moved to California in the 1860s. His father, Frederick, manufactured special cements so he had become familiar with concrete before leaving England. In the 1880s he patented several improvements to reinforced concrete construction, which included systems of expansion joints and the use of twisted iron bars to increase bond. Ransome began building in 1884 by erecting either part or whole structures mainly of reinforced concrete. The first of these were the Arctic Oil Works at San Francisco and a complete reinforced concrete mill building in 1885 for Starr and Company at Wheatport in California. In 1889 he built the first reinforced concrete bridge in the United States. After forming the Ransome Engineering Company to undertake his many commissions throughout the country, he continued to invent techniques and processes in reinforced concrete construction. One of these was a patent he took out in 1891, and again in 1894, for the glass lens of a prism set into concrete for lighting basements. Ransome and several others in the United States were greatly admired among European architects for their large factories. Walter Gropius, Le Corbusier, Erich Mendelsohn and others were greatly influenced by the simplicity of line, large windows, and the symmetry of repetitive elements. The functional form of the reinforced concrete grain silos in North America also provided inspiration.

Another Frenchman who made a unique contribution to the understanding of reinforced concrete was Auguste Perret (1874–1954). His work was said to be the first truly rational and efficient expression of concrete in the world. Perret's buildings expressed a classical influence in their design, although he eschewed the Renaissance movement and the teachings of the Ecole des Beaux-Arts in Paris. He had a fondness for the column as a source of strength and beauty in design, lavishing much attention on it in his works. In 1922 his church of Nôtre Dame du Raincy produced universal admiration, as it came to be regarded as the most revolutionary design in the first quarter of the twentieth century. Here Perret used tall, round, tapering columns in four rows with vaulted slabs and large areas of glazed non-loadbearing external walls. His work is considered to have reached a peak about 1927 with a general criticism thereafter that it expressed too much classicism.

The development of reinforced concrete received a considerable impetus with the work of Robert Maillart (1872–1940), the renowned Swiss engineer and pupil of Hennebique. Indeed no one had exploited so effectively the combination of structural and aesthetic possibilities of this material. Less bound by conventional formulae than his contemporaries he was not afraid to use his intuition as well as his very great technical knowledge. His impact on engineers and architects was most pronounced in his bridge designs, but he also had new ideas on building construction.

Maillart took the beamless floor slab developed by C.A.P. Turner in the United States in 1908 and perfected the mushroom column with flat slab to a fine state. Turner's beamless floor slab saw its best-known use in the Van Nelle factory in Rotterdam, built in 1927–28 to the design of J.A. Brinkman and L.C. Van de Vlugt. Maillart's first example of his beamless floor slab on point supports was the warehouse in Zurich-Giesshubel of 1910. Here the columns merged their hyperbolic forms into the floor slab without the drop slab used by Turner. Later the curved top of the column gave way to straight splays in the form of an inverted cone. In his bridges Robert Maillart developed the stiffened slab-arch, which acted structurally like a suspension bridge in reverse. Amazingly thin arch crowns were achieved. He also used the three-hinged arch with integrated deck slabs for the larger span bridges and then provided variations on this principle.

The other great structural engineer of the early twentieth century was Pier Luigi Nervi, born in 1891 in Italy. His first work of note was the Stadium Communale in Florence (1930–32) which was greatly admired by architects. At the end of the Second World War in Italy, I had the pleasure of watching an army football match played in this splendid arena.

The next development to have a profound influence on reinforced concrete design and construction was prestressed concrete. Although we regard this form of concrete today as a post-Second World War technological breakthrough, it is interesting that its origins go back to 1886. At this time C.E. Dochring, a German builder, considered that a form of pretensioned iron wires in concrete would provide an initial compressive stress. He experimented with small flooring units, but was unsuccessful because he did not have high tensile steel, nor was his concrete sufficiently high grade — two factors that are essential in prestressing.

Early in this century, research had been carried out by Stussi and Whitney and others but it was Eugene Freyssinet, the French engineer (1879–1962), who made the most significant contribution by eliminating tensile stresses. This was done by stretching the reinforcement so that, upon release, it would impose a compressive stress over the whole concrete section throughout the life of the structure, and over a wide range of loadings, so as to avoid cracking under working conditions. Freyssinet's experiments began in 1926 and continued for several years. The most effective proof of his technique was his successful arresting of the sinking Gare Maritime at Le Havre in 1934. However, in 1929 the Public Works Department in New Zealand used pretensioned No. 8 wires in concrete fence posts after roads were realigned in the early stages of constructing the Waitaki Dam. (The further development of prestressing is outside the scope of this book.)

It is fitting that in our earthquake-prone country the first book to be written in the English language on earthquake resistant buildings was by a New Zealander. In 1926 Charles Reginald Ford, a leading architect and structural engineer, had published *Earthquakes and Building Construction*, a work based on evidence of damage caused by such earthquakes as those in San Francisco in 1906, Tokyo in 1923, and Santa Barbara in 1925.

A more technical, design-oriented book was *Earthquake Resistant Buildings* by another Aucklander, S. Irwin Crookes Jr, published in 1940. When I attended the School of Architecture it was a set textbook for lectures in reinforced concrete construction delivered by 'Sammy' Crookes.

Mixing concrete

The Frenchman R. Foret, in 1892, established from his research into concrete strengths, using different quantities of water in the mix, that higher strengths are obtained by keeping the mix as dry as possible. There was the problem of proper placing of the mix and the labour of intensive ramming — wet mixes were so much easier to handle.

For many years the only way of mixing concrete was by hand; that is, by turning it over with a long-handled shovel on a board platform until the coarse aggregate (gravel), fine aggregate (sand), cement and addition of water resulted in a workable mix. The mechanical concrete mixer first appeared about 1847 but presumably it wasn't particularly successful. In

1854 Louis Cézanne invented a hand-operated mixer and this basic type can sometimes be seen today. Improvements followed until effective mechanically powered mixers came into use.

Electrically driven concrete mixers appeared in New Zealand early in the twentieth century. In 1908 the engineer and building contractor C. Fleming Macdonald scored a first for Dunedin when he used electrically operated mixers and also hoisting gear in that city for the construction of the New Zealand Express Company building in Bond Street.

Some very large engineering projects necessitated centrally located batch-mixing plants on site where the materials were measured and fed into a large mixer. This method was used successfully at Ohakea and Whenuapai in 1942 during the construction of heavy-duty concrete runways designed as 'tongue and grooved' hexagons. They were intended to take Flying Fortress and Liberator bombers. However, the Battle of the Coral Sea averted the need for such operations. After the attack on Pearl Harbor and consequent mobilisation, I was seconded from my Army engineer unit to spend several months on site at Ohakea with one or two other Public Works Department cadets. One of my duties was to test the concrete using test blocks formed in the standard 0.3-m-high brass cylinders and, when sufficiently cured, putting them in the testing machine to determine the compressive strengths.

Another task was to check each morning on the number of hexagons of concrete that had been poured in the previous 24-hour period. Shifts were worked between 6 a.m. and the following 2 a.m. and each of the 60,000 hexagons of concrete was marked on a plan of the runway more than 8 m in length. In addition there was the plan of the taxi strips and handling these plans outside in windy conditions was no easy task.

The next development on the world scene was the use of ready-mixed concrete conveyed by specially designed trucks in which a slowly rotating drum kept the concrete mix agitated until it could be poured from the vehicle down a chute into the formwork. The ready-mix concrete truck first appeared about 1926 in the United States and is in universal use today.

Chapter Two
The early days in New Zealand

1840–59

There does not appear to be a record of any claim for the first concrete structure in New Zealand. What we do know is that concrete as a construction material was introduced very early in the development of the young colony and found considerable favour throughout the latter third of the nineteenth century. Official records in the Blue Books show that as early as 1843, 158 casks of cement from England were imported through the Port of Wellington. Returns are incomplete for part of 1844 but record 89 casks of cement. There were no imports of this commodity listed for 1845 and no Blue Book for 1846. In 1847 there were 20 casks from Great Britain — that is, from England, as no other part of Britain manufactured cement at that time — and 24 casks from 'Elsewhere' totalling 44 casks, although the separate total given was 20 casks. Some poor arithmetic and some confusion over the origins probably account for the discrepancy. England was the only country exporting cement in 1847 so probably these 24 casks were imported through Sydney, having originated in England. It is fairly certain that such small quantities of cement were used for small engineering or military purposes and not for monolithic buildings. The records do not state whether these early imports were of Roman or Portland cement. Portland cement was being imported by 1862 and could well have been brought in much earlier, although it was not used much in Britain until the 1860s.

An early brief account of so-called concrete appeared in the *Lyttelton Times* on 17 April 1852: 'A neat cottage is nearly completed, built of concrete, a mixture of small gravel, sand and quicklime, in the respective proportions, I believe, of 4, 3, and 1 parts. There is also a small kitchen and flower garden, with a few fruit trees.'

The earliest evidence I have seen of concrete, still visible, is a rather crude retaining wall at Fyffe House in Kaikoura. In August 1857 the whaler George Fyffe engaged two men to excavate for a store shed and during their labours they discovered a large moa egg. The report in the *Lyttelton Times* of this find is evidence of the date of Fyffe's shed, which has long disappeared. The concrete wall is against a bank only a few metres from where he built a timber cottage. Early photographs show the shed hard against the bank so there is every reason to believe the concrete wall dates from this time.

One of the earliest recorded uses of concrete bridge piers and abutments was in 1859 on the outskirts of New Plymouth. The second bridge over the Waiwakaiho River, built in 1867, had two piers; the superstructure was of puriri, with iron rods and bolts to form a bowstring arch truss. This structure was replaced in 1907 by a reinforced concrete bridge, described in Chapter Seven. Today all that remains of the concrete-filled cast-iron cylinder piers of the 1867 bridge can be seen on the banks. They are 1.2 m in diameter. A flood in 1867 washed the truss (an 1859 structure) downstream, but it was reused and the bridge was extended by two shore spans.

The 1860s

The pioneer colonial example of a complete monolithic building is the two-storey house built in 1862 by John Gow near Mosgiel. An early settler engaged in farming on the Taieri Plain, he had built previously on his property, Invermay, two modest homes.

MUSTERERS' QUARTERS, LAKE COLERIDGE STATION, CANTERBURY.
Said to have been built in 1861, and certainly standing in the 1870s, this building is still used on occasions.

The first, a sod cottage, was followed by an adobe (mud brick) house before he decided on the more substantial structure that still stands today. Its nine rooms included a billiard room. Its concrete walls are plastered on both faces, and it has a slate roof. In 1956 the University of Otago purchased the Invermay property for use as a research centre, with the house being occupied as staff living quarters. This building ranks as the oldest extant concrete building in New Zealand and by world standards it is a veteran. It would be interesting to know what prompted John Gow to decide on concrete as the material for his house.

Although its date has not been firmly established, there is a small concrete building on Lake Coleridge station in Mid-Canterbury that is said to have been built in 1861. These musterers' quarters are an L-shape with a roof of corrugated iron. The concrete walls are in remarkably good condition today. The present owners cannot throw any light on its origins other than the fact it was there in 1878 when their grandfather acquired the adjoining Acheron Bank station. It must have been a remote place at that time for such an innovative building material. The original portion of the old homestead, not far away, is cob, so possibly there was a fear of fire and consequent desire to use fire-resistant materials. There may also have been a shortage of timber in the district.

An early Taieri settler was Alexander Campbell who took up land about 4 km north of Outram. In 1863 he was joined by his 19-year-old brother Robert and they lived in a sod hut at first. Before long they built a stable for their draughthorses, using concrete with 250-mm-thick walls. Robert had a book on how to build in concrete and this may have been the spur for deciding on this material. The stable is now a general-purpose farm shed still in very good order. A concrete house constructed close by in the 1890s is described later.

An early West Coast example of concrete foundations in a modest timber structure is the former Donovan's Store at Okarito. This dates from c. 1866 and was originally the Club Hotel. It is being restored by its present owner, the Department of Conservation.

The large cowshed at Tarureka in the Wairarapa town of Featherston, built in 1868 for James Donald, had concrete floors with two channels between the double row of bails. There was a constant supply of running water in these to ensure cleanliness. The yards adjoining were also concreted, showing a noteworthy early use of this material in such situations.

In late 1864 the foundations were begun for the Anglican Cathedral in Christchurch and completed in the following year. To achieve a depth of 1.8 m under the nave walls and 2.1 m under the tower, a total of 2,000 barrels of cement was used. The architect, George Gilbert Scott (later knighted), remained in England but he sent out Robert Speechley, also an architect, as his supervisor.

The New Zealand and Australian Land Company, active in South Canterbury and Otago last century, was noted for its progressive ideas on farming. This did not extend to erecting monolithic concrete until the late 1870s. On its Totara Estate, a few kilometres south of Oamaru, the readily available limestone was used for the manager's house and farm buildings. The floor of the barn, however, is concrete and it was probably built in 1868. This and its adjacent buildings were restored by the New Zealand Historic Places Trust in readiness for the centennial celebrations to mark the first shipment of frozen meat to England in February 1882. The slaughterhouse for the preparation of the carcases was erected in stone, timber and galvanised iron but the only portion remaining is part of a stone wall and the concrete floor. It was said that the smooth surface of the latter was obtained by the addition of blood to the concrete mix. A gutter formed in the concrete and covered by a grating allowed the blood from the animals to run into a trough in the adjoining pig shed.

Invermay, near Mosgiel.
Built as a farmhouse in 1862 by John Gow, Invermay has nine rooms including a former billiard room. It is now owned by the University of Otago and is the oldest extant concrete building in New Zealand.

Chapter Three
The 1870s

This decade saw a remarkable number of concrete structures erected in New Zealand, many of which still survive. About this time there was considerable interest in England, the home of early concrete development, expressed in a wide variety of concrete structures virtually all of plain concrete, there being only isolated attempts to use reinforcement with iron in some form. However, one would not have expected such a small and young colony, a veritable outpost of the Empire and remote from technological developments, to be so adventurous with a new material for building. Perhaps these apparent drawbacks acted as a spur to trying out new ideas. In addition skilled labour was generally in short supply for traditional building trades such as carpentry, masonry and bricklaying. In the drier country districts especially, many settlers had tried their hand at building their own homes with earth materials such as sod, adobe, *pisé de terre*, wattle and daub, and the more common cob. It may not have been such a big step to try mixing one's own concrete and erecting the necessary formwork. Certainly there were several innovative architects and engineers in New Zealand at this time who were keen to see the properties of concrete advanced in actual structures. Where good building stone was unobtainable or scarce the new material seemed to have much to offer for its resistance to both fire and earthquake. Another advantage to be exploited later was its plasticity, which allowed it to take any desired form.

John Logan Campbell, known as the 'Father of Auckland', had seen concrete buildings in England during a prolonged stay there, and at the 1870 Paris Exhibition formwork for concrete was on display. He was sufficiently impressed to have some sent out to New Zealand and these were used for concrete additions to his house, Logan Bank, in Auckland. This work took place in early 1871 but the building is no longer standing.

Another of Auckland's leading citizens also became enamoured of the new material and in 1871 he made a substantial concrete addition to his timber house in Epsom. Josiah Clifton Firth was co-founder of a large flour mill in the city, and he also had an extensive tract of land at Matamata, which he was developing in a most progressive manner. His house, Clifton, was of two storeys and had been purchased from a previous owner. In order to provide additional and better accommodation for his family he altered the house and built a square concrete tower of four storeys, after demolishing the old kitchen to provide a link. It is some 15 m high and crowned by battlements which have deeply recessed arches beneath them. On the corners are paired heads of rams, lions and Maori chiefs. This top section oversails and has a decorative corbelled cornice below. Windows have traditional label moulds. The uppermost floor was used as a family museum but the flat concrete roof leaked in heavy rain. It is recorded that Chinese ginger bowls were pressed into service to catch the drips! Below were two bedroom floors and at ground level, a laundry. It was 1873 before all the work was finished.

CLIFTON, EPSOM, AUCKLAND.
The concrete tower was begun in 1871 by J.C. Firth as an addition to an existing timber house. It is embellished with quirky decorative features on the exterior.

MEN'S QUARTERS AND IMPLEMENT SHED, ABBOTSFORD, OUTRAM.
Built in 1870, the block is a part of the large steading designed by Mason and Wales of Dunedin. A larger building opposite, with stables and, originally, cattle stalls was converted a few years later to a shearing shed.

By the 1870s the use of concrete was being developed for structures other than buildings. At Langley Dale station in Marlborough a small concrete scab-dip was constructed, probably in 1870 when considerable quantities of tobacco were first grown. This dip, still standing today, is a reminder of the widespread curse of scab disease in the sheep flocks of the colony — an affliction that persisted for many years.

William Adams was a pioneer farmer who took up his lower Wairau run in 1857 and two years later became the first Superintendent of Marlborough Province. He planted his own crop of tobacco as a means of providing a solution in which to dip the sheep. This was prepared and boiled in two large iron tanks encased in concrete and supported on a base of schist rock and brick. When ready, the tobacco solution was led to the nearby concrete pot dip — a well almost 1.8 m deep from which the thoroughly immersed sheep scrambled up a concrete race leading to the stone-paved draining yard. A top-hung gate with a tipping device allowed the sheep to slide into the dip from a small pen. Water for the operation came from a stream alongside the iron tanks.

Probably the earliest concrete farm buildings of any consequence were those designed by the Dunedin architectural firm of Mason and Wales for James Shand in 1870. As a progressive farmer he no doubt considered the new material very suitable for his farm buildings at his Abbotsford property near Outram. There are three sections in the steading. A two-storey implement shed with men's living quarters upstairs forms one L-shaped block. Another unit comprises stables for twelve

STRATHCONAN, NEAR FAIRLIE.
Built in 1877, this portion was altered in 1926 to give a Californian bungalow overlay. The photograph does not show the rest of the extensive house, part of which is timber.

draughthorses with a harness room, three stalls for riding hacks and two loose boxes. At right angles to this block is a wing originally having 38 stalls for cattle, but a few years later this was redesigned by the architects to allow for a shearing board and catching pens. These buildings, displaying such an early profession of faith in concrete on a fairly large scale, deserve full recognition as pioneer structures. The most significant modification over recent years has been the large covered yards adjoining the shearing unit. James Shand built another concrete steading in 1881, a few kilometres distant on his Berkeley property near Henley.

In 1871 Donald McLean, who had been manager of the Levels station for the New Zealand and Australian Land Company, took up the Strathconan portion of the Albury run. He began building a substantial homestead about 3 km from Fairlie, using concrete for the single-storey construction. It was not completed until 1877 and was a pleasant-looking house with long verandahs.

A new owner took over in 1919. R. Allan Grant had rather different ideas on domestic design and set about making drastic alterations. Two concrete bearing walls were removed to form a large living and dining room having two large fireplaces of shell-bearing rock. Beams were exposed.

SUNNYSIDE MENTAL HOSPITAL, CHRISTCHURCH.
The original wing was designed by B.W. Mountfort and dates from 1871–74. It was extended by T. Cane and the Public Works Department.

BREAKWATER, OAMARU HARBOUR.
Designed by John McGregor this breakwater was begun in 1872 but was not completed until 1884. It was a very costly project for such a small town, and became increasingly so in 1886 when storm damage necessitated the replacement of 81 m of breakwater.

The house was given a Californian bungalow overlay emphasised by the low verandah walls, but in this case covered with river-run shingle as decoration. These alterations took place between 1920 and 1926. There are also parts of the house in timber. Today Strathconan has much the same form and was still structurally sound at the time of writing.

Allan Grant had extensive gardens laid out in a different arrangement than hitherto. Along the north side there are terraces with lengthy pergolas wreathed in wisteria. Considered to have been somewhat eccentric, he built a small circular Roman temple in wood to house a bowser (petrol pump). Today it is a vintage item as well as a delightful folly.

The first really large essay in concrete construction for a public building was the West Wing of Sunnyside Mental Hospital (then known as Sunnyside Lunatic Asylum) in Christchurch. The architect, Benjamin Woolfield Mountfort, who was the Canterbury Provincial Architect, used Portland cement concrete for the 30 m by 9 m building of two storeys, because of the fear of fire in such an institution. Construction began in 1871 but was halted for a year because of a shortage of suitable tradesmen. Until then the work had been largely carried out by patients and men applying to the government for charitable aid. The wing was completed in 1874 and consisted of four wards for women patients. The subsequent stages were designed by Thomas Cane and the Public Works Department (PWD). After a disastrous fire in 1888 had destroyed the roof and upper storey of the West Wing, reconstruction and modifications were carried out by the PWD. Instead of concrete, brick was used with plaster finish. Evidence of this can be seen in the now disused building. Relatively modest in scale by comparison with Cane's design, the West Wing displays Mountfort's skill in simple but careful detailing of the Gothic-inspired design he chose in an attempt to avoid an institutional atmosphere.

A major engineering work at this time was the construction of the breakwater begun in 1872 to protect the small harbour at Oamaru. The engineer was John McGregor, who saw the first block laid on 10 September of that year. It was not completed until 1884 when it reached a length of 564 m at a cost of £156,000 — a very large sum of money in those days. The breakwater consisted of two rows of concrete blocks each weighing about 25 tonnes. There were cross ties to form pockets filled with rubble from an adjacent quarry. The top surface was of concrete, about 2 m above extreme high tide. In August 1886 a tremendous storm seriously damaged the breakwater. Two large gaps of about 21 m and 61 m were torn out so that repairs became a major undertaking in that year. The financial burden was such that by 1894 the Oamaru Harbour

House in Thorndon, Wellington.
This 1874 house was designed by W.H. Clayton, Colonial Architect, for his family. It is built of lime concrete and survived until the 1980s when it was demolished to make way for Queen Margaret College teaching needs.

Board was in receivership. It was not until 1936 that the breakwater was finally completed, but this included raising it by another 2 m, using stone from the board's quarry nearby.

Supplies of potable water for towns is an essential public utility demanded of urban councils. In 1870 the capital city, Wellington, was still reliant on wells and domestic rainwater tanks. In that year when asked to test samples Dr James Hector, the government's adviser on scientific matters of all kinds, condemned them as quite unsuitable. The favoured site for a reservoir was the Upper Kaiwarra Stream, but this was occupied by a number of goldmining companies, some of which were active. To clear the way for the city council to acquire the necessary land, the Wellington Waterworks Act was passed, thus enabling the construction of the city's waterworks to proceed. In November 1873 the Karori Reservoir came into use.

The valve tower is a visual delight — octagonal in plan, its Gothic-mannered turreted timber superstructure is perched on a concrete shaft that is flared at its top over scroll corbels. During normal operating levels only this upper portion can be seen. The reservoir, with an earth dam, is lined with concrete. Some 37 years afterwards a larger upper dam was built of concrete, and this is mentioned in a later chapter.

Portland cement was used extensively, although not as concrete, in the construction of the graving dock at Port Chalmers. Begun *c.* 1869 and completed in 1872, the dock absorbed some 14,000 tonnes of Portland cement in the mortar for jointing the locally quarried breccia to form a structure 100 m in length and 15.2 m wide with stepped sides or altars. The engineer was Robert Hay of Dunedin. This superb example of Victorian building construction still exists, but regrettably cannot be seen for it has been filled in to provide container storage space for the Port of Otago. What a splendid discovery this would be for some future archaeologist intent on exhuming our early industrial history.

In the capital city, William Henry Clayton, Colonial Architect, decided to build his own house in concrete. It was a two-storey building in Hobson Street, Thorndon, on land leased from the hospital trustees. Tenders were called in June 1874 and the house was built soon afterwards. Although Clayton described it as a cottage in his advertisements for tenders, there were twelve rooms. It was not built of Portland cement concrete but used lime, clay and coarse aggregate. This house had the distinction of being the first in Wellington to have hot and cold water laid on. It became part of Queen Margaret College and

St James' Presbyterian Church, Auckland.
The interior has good natural lighting and a gallery that was added later.

Opposite: Valve tower, Karori Reservoir, Wellington.
This attractive structure has a concrete shaft. The reservoir, which came into use in 1873, has a concrete lining, although the lower dam is puddled clay.

was drastically altered over the years to meet teaching needs. Recently it was demolished to make way for new accommodation.

A remarkably early use of concrete for a place of worship is the very fine church in Beresford Street, Auckland, built for the Congregationalists. Construction began in 1874 to the design of Philip Herapath, an architect well known for his city buildings, but it was 1876 before the church opened for worship. Herapath was the first architect in Auckland to design a public building in monolithic concrete and it is a tribute to his skill that this church is still standing. It is possibly the oldest concrete church in existence. The Congregationalists later united with the Presbyterians, and since September 1964 it has been known as St James' Presbyterian Church. The front elevation has an impressive Greek Doric portico.

The interior is quite splendid, with its Early Victorian character and full length galleries (added in 1881) contributing to its appearance. The acoustics are reputed to be superior, no doubt assisted by the gently curved rear wall (behind the colonnaded entrance). Seating in the form of low pews is arranged in an arc to face the minister more comfortably. A flat ceiling is supported by a trussed roof and is divided into decorative panels to give interest. Both side elevations have tall windows giving a very good quality of natural light. Unfortunately the building is being damaged by the vibrations of heavy traffic on the adjacent motorway. The high cost of strengthening is beyond the resources of the small congregation and the church is likely to be demolished in spite of being a listed building.

One of the earlier uses of concrete in a public building, although in a limited manner, was in the new courthouse at Lawrence. Tenders were called in January 1874 for the erection in stone and concrete to the design of the Dunedin architect, David Ross. The concrete portion consists of a 16.7-m-long colonnade with a slab roof. The rest of the building is rendered brick. The *Tuapeka Press* described the local reaction to the use of concrete in December 1874: *(To page 38)*

COURTHOUSE, LAWRENCE. Built in 1875 of plastered brick, the courthouse has its concrete portion in the colonnade. The contractor had difficulties with the concrete construction and blamed the architect, D. Ross of Dunedin.

ST JAMES' PRESBYTERIAN CHURCH, AUCKLAND.
Designed by Philip Herapath, St James' opened in 1876 as the Beresford Street Congregational Church. It is a pioneer in concrete church construction and at the time of writing was still in use.

THE ELMS, KAHUTARA, KAIKOURA DISTRICT.
The Elms was a large estate with a grand concrete house built in 1875 for the Bullen brothers. The cottage is in two separate units, and was built for the gardener and the coachman *c.* 1875.

BLACKSMITH'S SHOP, CLARENCE RESERVE, KAIKOURA DISTRICT.
This small 1870s smithy has only the one window above the door. It was not used for farriery work.

> The supports of the concrete roof of the Colonnade of the new Court House fell to the ground yesterday morning as soon as the supports and wood framing were removed from underneath it. The noise of its falling was heard all over Lawrence. A great misfortune to the Contractor who had the building in an advanced state. The rebuilding will not take very long and previous experience, no doubt, will be a guide to prevent a similar occurrence.

The story continued on 10 March 1875:

> The supports of the concrete roof of the Colonnade of the Lawrence Court House are expected to be removed this week in the presence of the Architect. It is expected there will be a large number present during the operation as grave doubts were entertained as to its stability. The concrete mass which forms the roof of the Colonnade is 55 feet [16.5 m] long and 5 feet [1.5 m] wide and has a thickness of 10 inches to 7 inches [250 mm to 175 mm], thus giving an arch of only 3 inches [75 mm] and altogether it must weigh 40 tons [40.6 tonnes].

And on 13 March 1875 the paper reported further:

> The Contractor for the Lawrence Court House is, figuratively speaking, on the horns of a dilemma. The Architect has been and gone and the wooden supports of the concrete roof of the Colonnade remain unmoved. The Architect, it appears, will not take the responsibility of giving instructions to remove the supports and the Contractor is afraid to remove them on his own responsibility in case he is placed in the same predicament as before and has to stand the brunt of re-erection. We do not profess to know anything of the capability of concrete, but an arch of 3 inches as in the roof which has to carry a weight of 40 tons, seems rather little and probably an experiment on the part of the Architect.

The Lawrence courthouse is still standing in a sound state.

In late 1874 construction began on the Ocean View Hotel in South Dunedin with a main frontage of 16.45 m. When it opened in the following year it was hailed as the first two-storey concrete building in the city as well as being its first hotel in this material. It was noted for its large assembly room available for meetings of local groups — a practice quite common at that time. The design was by N.Y.A. Wales of the Dunedin firm of Mason and Wales, already quite experienced in concrete construction. This hotel is still in use.

An interesting connection exists between a South Island property and the renowned designer of the Crystal Palace in London for the Great Exhibition of 1851. When in 1874 John Reid built a splendid house at his Elderslie estate in North Otago, he had the extensive grounds laid out to a plan prepared by Sir Joseph Paxton. The first part of the property was purchased in 1865 and Reid sent gardeners to work on the layout from this time. As Paxton died in this same year, Reid may have obtained the plan before buying the land. It is possible that Sir Joseph Paxton designed the little bridge over a stream near the main gates. It has concrete piers and abutments. Today the grounds are derelict as a consequence of the house having been destroyed by fire in 1957.

In the district just south of Kaikoura several sheep runs had concrete buildings erected in the 1870s. Between the Kahutara and Kowhai Rivers, George Francis Bullen and his brother Frederick built a substantial two-storey house which they named The Elms. The builder was John Alves of Dunedin. There are eighteen rooms and on its completion in 1875 there was a grand house-warming to mark a truly social occasion. The surrounding garden was spacious with many trees that in time provided a fine setting. To the rear is a concrete barn, formerly used as a granary, and close by are the stables, likewise of concrete. The stables building is now a woolshed and has been modified externally. The construction consists of no-fines concrete walls using a mix of river shingle and Portland cement with little or no sand so that a rather porous texture is produced. Waterproofing is provided by rendering in cement plaster on both faces. Today there is evidence of severe cracking possibly caused by earthquakes over the years. Presumably the homestead is also of no-fines concrete.

At the roadside, now State Highway 1, is a concrete house built as flats for the coachman and gardener. This building has a steep pitched corrugated iron roof with two squat chimneys abutting each end of the short ridge. Apart from the closing in of a small corner verandah and one at the rear there has been little external alteration over the years. The exterior face has been rendered and lined as very large blocks to break up the otherwise finely textured surface. Both The Elms homestead and the granary are in very strict private ownership and could not be inspected.

The Bullens also owned the Greenhills run near Clarence Reserve on which they built, in the 1870s, a large single-room concrete bunkhouse and cookshop for the shearers. It had tiered bunks to sleep more than 70 men and there was an open fireplace spanning an end wall. The shearers sat on forms on either side of a long table in the centre of the room. During

BLACKSMITH'S SHOP, SWYNCOMBE STATION, KAIKOURA DISTRICT. Built for the Keene brothers, this was one of a number of concrete farm buildings erected in the district during the 1870s.

shearing there would be at least 100 men to be fed in what must have been very cramped quarters.

In the same district is Swyncombe station, formerly owned by the Keene brothers and their father, who remained in England. They also had the Greenhills run prior to the Bullens. Swyncombe has a concrete blacksmith's shop dating from the 1870s. Today, although empty of smithy equipment and unused except for hay storage, it retains its character with segmental headed windows giving it a rather more refined appearance than is customary. Swyncombe station was the first in the country to introduce mechanical shearing when Wolseley machines were installed in 1884 by the owners, the Wood brothers.

Yet another concrete farm building of the 1870s is still standing at Clarence Reserve station, a few kilometres further along the Inland Kaikoura route. A small blacksmith's shop, it is a rather forbidding boxlike structure without windows. The presence of all these buildings in this district at this time suggests that concrete was considered a very suitable material for farm structures and one to be encouraged, although there did not seem to be any general trend elsewhere as a result.

It would seem, too, that concrete was regarded as an ideal material for an explosives magazine. One was constructed in Lyttelton in 1874 for civil purposes. It was later taken over by the military and is still standing.

In 1876 a small farmhouse was built in Mercer Road, in the Waitepeka district, about 10 km from Balclutha. In the following year it was occupied by Archibald Mercer and his family. Exterior walls are concrete, 235 mm thick, and the interior walls are 150 mm thick, all being in remarkably good condition. The house is gabled and roofed in corrugated galvanised iron with dormers and gable-end windows in the upper floor. On the south side there is a timber lean-to and the rear has a small service wing, also of timber. Along the front elevation is a full length verandah which has a gablet over the centrally placed entrance with decorative bargeboards. Cast iron valances between the posts may have been added a little later. The two chimneys penetrating the ridge are concrete, which was not usual at that time.

This delightful little house has been retained in a commendably intact state and is an excellent example of new

FARMHOUSE IN MERCER ROAD, WAITEPEKA, SOUTH OTAGO.
Built in 1876 by Archibald Mercer for his family, the farmhouse is now back in the ownership of family descendants and the original cottage is being faithfully restored. Sympathetic additions have been planned in place of the crude lean-tos.

GOLDIES BRAE, WADESTOWN, WELLINGTON.
Popularly known as 'The Banana House', Goldies Brae was designed by the owner, Dr Alexander Johnston, to make use of solar heating through the fully glazed crescent-shaped verandah. It was built in 1876.

technology being adopted in a rural area. A short distance away is another concrete cottage, smaller and with an attic bedroom. It is now quite derelict without any roof covering and is badly cracked from the misguided efforts of a bulldozer operator. Nevertheless it appears to have been a well-built and comfortable home with the walls being plastered on both faces and an integral colour used on the exterior.

About 1875–76 Charles Meyer, the owner of Blue Cliffs station in South Canterbury, had concrete additions made to the timber homestead of 1866 built for his predecessor, John Hayhurst. Meyer included a drawing room, dining room, hall, a separate kitchen and two bedrooms. A new cookshop and men's quarters was also built in concrete at this time. However, neither of these structures exists today.

A pioneer house of concrete in Wellington is that built in 1876 in Wadestown and known originally as Goldies Brae. The owner was Dr Alexander Johnston. Because of its shallow crescent shape it has become popularly styled 'The Banana House'. The walls are 300 mm thick, enclosing ten rooms all on one level and facing onto a conservatory-cum-gallery with an encaustic tiled floor. The roof is slate. Revolutionary in its form, this house is noteworthy for its use of solar heating through the conservatory facing the sun. Dr Johnston conceived the design himself and was in effect his own architect. It is a surprising piece of architecture and rightly has a high preservation classification from the New Zealand Historic Places Trust.

The Wanganui Racing Club's grandstand, built of timber in 1876, still features an extensive and gently stepped concrete apron as access.

In Oamaru the New Zealand Loan and Mercantile Agency Company Limited had a building erected in concrete, *c.* 1873 and possibly in 1876, in Harbour Street. It was a warehouse designed by Robert Forrest and David McGill of Dunedin but was rather small, being replaced in 1882 by the present large three-storey stone structure.

The renowned Christchurch architect, Benjamin Woolfield Mountfort, designed many churches for the Anglican Diocese of Canterbury. Having come from England he had an understandable preference for stone, although he also used brick and timber. Mountfort was not regarded as an advocate for concrete in ecclesiastical building, so it is interesting to observe his use of this material in the foundations and lower walls of the Church of St John the Baptist in Rangiora.

While in partnership with his brother-in-law, Isaac Luck, he had designed a small church of pitsawn timber which opened for worship in 1860 on the site in High Street. Although it was enlarged in 1864 it soon became necessary to consider building a bigger church and Mountfort was commissioned to do this. To allow services to continue without vacating the existing church, it was decided to build the new around the old. The foundation stone was laid on 4 November 1875. First

CHURCH OF ST JOHN THE BAPTIST, RANGIORA. Building began in 1875 round the earlier church to enable services to continue as long as possible. The architect, B.W. Mountfort, used part-height concrete walls.

the chancel was removed to the cemetery for use as a chapel, and the transepts were dismantled to be rebuilt in the new design. These have concrete walls about 2.5 m high, with the old timber walls above. The nave was then built with concrete walls up to windowsill height so that the older portion could be removed. However, it was not until 1882 that the church was completed. The Gothic treatment of the design uses vertical boarding with battens, and narrow lancet windows. The roof is timber framed and clad in corrugated galvanised iron. The use of these low concrete walls is certainly unusual, but the design has been handled skilfully and their use is logical for the manner in which the work was carried out. St John's Church has a fine interior and a very attractive setting on a corner site where large trees enhance the building.

The construction of the renowned Rimutaka railway over steep hills in the 1870s entailed some use of concrete, including the Siberia water tunnel and its dramatic intake structure. The intake was perforated to allow surface water to drain into it and be diverted from the railway line. Subsequently it was raised as shingle began to fill up the gully behind the railway embankment. This work was completed in 1877.

At the Cross Creek terminus of the Rimutaka Incline, famous for its use of a centre rail gripped by the Fell locomotives to cope with the 1 in 5 grade, there were various buildings. One was the locomotive shed, which had concrete foundations, and there was a concrete ash pit and foundations for a turntable. Today all these relics can be seen, as the railway line has been diverted since 1955 by a major tunnel to remove any appreciable grade, and now the public can use the old track as a walkway.

One of the early architect exponents of concrete was Francis William Petre who probably did more than any other architect of the period to further interest in this material. Judge Henry Chapman commissioned him to design a substantial house in Dunedin. This home, Woodside, was built to take advantage of views of the city, bay, and bush surroundings. The following description of it appeared in the *Saturday Advertiser* on 5 February 1876: *(To page 48)*

St Dominic's Priory, Dunedin.
Designed by F.W. Petre and built in 1877, this is a remarkable multi-storey building for its time. The harmony achieved by the arrangement of windows is pleasing and logical for early concrete construction.

WOODSIDE, LOVELOCK AVENUE, DUNEDIN.
Woodside was designed by F.W. Petre for Judge Chapman and was completed in 1876. Noted for its splendid interior, it was highly regarded for its time. Today it is known as Castlamore.

GLENMARK HOMESTEAD, NORTH CANTERBURY.
This was a very large, Gothic-mannered mansion built 1877–81 to the design of S.C. Farr for G.H. Moore. The unusual construction had rusticated weatherboards and interior linings of timber panelling. It was destroyed by fire in 1891. The photograph shows part of the plain concrete servants' quarters at one end.

The Cliffs, St Clair, Dunedin.
Designed by F.W. Petre, The Cliffs was completed in 1876. Popularly styled as 'Cargill's Castle', the rather Italianate building has been a picturesque ruin for many years.

One of the surest indications of the prosperity of a young country is the growth of a refined taste in architecture. In the early settlement of a colony, people are not over particular as to the sort of dwellings which they live in: but as things calm down, and the excitement attendant on the founding of a new settlement ceases, sensible folk look out for comfortable homes. In this respect the settlers of Otago have not been behind their neighbours. The cosy snuggeries and fine mansions which are springing up around us every day serve to illustrate the progress which we are making in the march of civilisation and refinement. We have been induced to make these remarks by a visit which we paid a few days ago to Woodside, the new residence of our respected fellow-citizen, Judge Chapman . . . The building is of concrete and the style of architecture is the Tudor Renaissance. The castellated gables and parapets which ornament the exterior of the structure give it a rather imposing appearance, and the octagonal chimneys and tower windows are in strict keeping with the entire design. As illustrating the superiority of concrete over other building materials, we may state that one wall which runs to the extreme length of 37 feet [11.1 m] is only 5 in [125 mm] thick, whilst the thickness of the strongest wall in the house is but 9 in [225 mm], yet not the slightest symptom of a crack is perceptible in any portion of the building. One of the principal features of the structure is the entrance hall with its tessellated pavement of fancy coloured encaustic tiles. In the centre of this pavement, the crest and coat of arms of the Judge's family are elegantly wrought in mosaic. The hall goes right up to the roof in height and it has a recessed gallery on three sides . . .

    In conclusion, we must compliment Judge Chapman on his taste and enterprise in erecting such a dwelling and too much praise cannot be bestowed on the architect Mr F.W. Petre for his skill and judgement in the designing of such a building. We understand Mr Petre is now engaged on the erection of a new house for Mr E.B. Cargill, which promises to become an ornament to the suburbs of Dunedin.

The house for Edward Bowes Cargill, also built of concrete, was sited high on a cliff top, and overlooking St Clair. Called The Cliffs by Cargill, it was popularly known as 'Cargill's Castle', the crenellations along the roof no doubt contributing to that image. For many years now the building has been in a ruinous state, with the elements playing havoc since roofs and windows disappeared. Cargill, the seventh son of the founder of Otago, Captain William Cargill, had five daughters of whom the eldest married F.W. Petre, which could account for the construction in concrete. The house was partly of two storeys with a tower and is somewhat Italianate in appearance. This is not surprising for Petre had a love affair with Italy, being consul for that country for many years. In its heyday The Cliffs was a substantial and well-appointed house. Cargill was a businessman with many commercial interests, including partnership with E.R. Anderson in the large Teviot sheep station in Central Otago. In 1897 he became Mayor of Dunedin.

Petre's masterpiece, so far as concrete essays are concerned, is the former Dominican Priory in Dunedin. Sited impressively on an elevated corner alongside St Joseph's Cathedral it is a superb building in its form, proportions and overall simplicity. Gothicism is expressed with modesty and refinement in the sharply pointed windows. There is no Gothic exuberance of glass-filled bays; rather, the solid walls are pierced with small windows as in Romanesque architecture — a logical design for the new material. One suspects that the avoidance of the curved Gothic arch in the windows was to simplify the formwork, as a four-storeyed building in concrete would test the ability of most contractors at that time.

The Priory was completed in 1877. The unreinforced concrete walls at ground level are immensely massive for this material, being 610 mm thick. The interior planning is noteworthy for two staircases. To give completely private access to the small infirmary, a totally enclosed doctor's staircase was built so that no one would be able to see him other than the patient and prioress. The main staircase is a fine structure at the south end of the building. It has been loosely described as a hanging stair in that it skirts the walls without any other structural support; however, it is in fact cantilevered in its ascension of four floors. Although the working drawings prepared by Petre do not indicate brickwork, a recent structural report states that only the first three bays are concrete, the rest being plastered brick and stone-faced brick. The stone portion was a later addition. Nevertheless this priory must rank as one of the finest extant concrete monolithic buildings for its period.

Of all the impressive homesteads built by runholders and station owners to replace earlier and more modest homes, it was the Gothic-styled mansion at Glenmark in the Waipara district in North Canterbury that set new standards of splendour. George Henry Moore had prospered with Glenmark by this time so he engaged Samuel Charles Farr, an architect in Christchurch, to design a fine house of two storeys with the best of materials and workmanship. Farr used concrete extensively. There was a continuous foundation and for much of the house he incorporated a most unusual form of construction. This consisted of concrete walls sheathed externally with rusticated weatherboards attached to a timber framework set in the concrete. On the inside faces there was a panelled lining kept clear of the concrete to allow air spaces in which the movement was controlled by ventilators. Against the hillside, at the rear and one end, was an enclosed courtyard with crenellated walls 4.5 m high. Opening into this were concrete rooms for kitchen and storage.

Begun in 1877 the house was not completed until 1881 or possibly 1882 but it had a remarkably short life. In January 1891 a fire destroyed it in an hour and today it is a stark ruin. Noted for his parsimony, Moore had refused to have repairs

4 BEACH ROAD, ASHBURTON.
S.C. Farr designed this house for his daughter and son-in-law, Dr McBean Stewart, with tenders called in 1877. It is much longer than the front elevation indicates and was completed in 1878.

made to a chimney cracked in an earthquake. It is believed this was the cause of the fire, which was accelerated by the blasts of air behind the panelling in the main rooms. In its brief heyday the house was noted for its magnificent setting with lakes and water gardens. Farr was responsible for designing the concrete weirs and control structures. As if these works were not enough, he also designed a bridge with concrete piers and abutments which is still in use, and the superb stable block, which is described in the next chapter.

The extent of concrete construction at Glenmark indicated that Farr had previous experience with this material. For a short period he was the Lyttelton Borough surveyor and in 1871 he introduced concrete street channels, at that time an innovation for the borough. In 1877 he called tenders for a single-storey concrete house at 4 Beach Road, Ashburton. This was the home for his daughter and son-in-law, Dr McBean Stewart, and it was completed in the following year. A long rectangular building, its short elevation faces the street and this presumably contained the surgery and waiting room. The symmetrical front has full-height bay windows and a centrally placed, segmental arched entrance. The corners have quoins formed in plaster. Today the house is in two flats.

A number of nineteenth-century architects took up land in New Zealand as did a few engineers. One was Edward George Wright, who settled on Windermere in Mid-Canterbury as a land grant in part payment for railway contract work. The homestead he built about 1877 is of two storeys with timber framing on concrete foundations. One portion of the house, containing the kitchen and boiler, is built of concrete. The garden has a detached conservatory with low concrete walls and floor. On the south side abutting the house a high concrete wall protects the garden, while behind the house a concrete service block accommodates a baker's oven, cool room and storage. Further away is the concrete coach house and stables with feed loft and groom's quarters upstairs.

No doubt Wright, as an engineer, had considerable faith in concrete and it is fairly certain that he engaged S.C. Farr to

SERVICE BLOCK, WINDERMERE, MID-CANTERBURY.
Sited behind the house this *c.* 1877 building had a baker's oven, cool room and store. The timber house has a concrete addition for the kitchen and boiler. The owner was E.G. Wright, a civil engineer who took up land in Mid-Canterbury.

STABLES, WINDERMERE, MID-CANTERBURY.
Distant from the house are the stables and coach house built *c.* 1877. The coachman had his room above.

design his buildings. Farr had already shown his interest in using concrete and much of the detail in the house points to his experienced hand.

Until the advent of reinforced concrete construction, bridge design made little use of concrete except for piers and abutments. An early combined road and rail bridge was designed by Harry Pasley Higginson, a consultant engineer, for the Waimakariri Gorge crossing where a ferry had been operating since 1872. Toll charges for this transport were sixpence for foot passengers and one shilling for horsemen. A loaded four-wheel dray cost the driver four shillings while sheep, pigs and goats were charged at a penny a head. In 1877 a fine bridge was completed over the fast flowing river, making some use of concrete in its construction. It has 2.4-m-deep wrought iron plate girder spans of 38 m, 33.5 m and 28.9 m. Two tall piers of rectangular section with rounded ends were constructed of wrought iron plates filled with concrete and set into rock to a depth of 0.3 m. The 32.9-m-high bridge also has substantial concrete abutments. Although concrete-filled cylindrical iron piers were not uncommon, this example was the first of such a

WAIMAKARIRI GORGE BRIDGE, MID-CANTERBURY. Completed in 1877 to the design of H.P. Higginson, this 32.9-m-high structure has the piers encased in cast iron plates — a common practice from this time for much smaller bridges.

Maze House, State Highway 8, Pleasant Point. This pleasing house, built in the 1870s, was designed by an engineer named Worthington for his own use. The verandah returns along the north side. It was acquired by the Maze family in 1883 and remained in their ownership until recently.

height in the country. By today's standards this timber-decked bridge is narrow, having been built for the railway which ran between Sheffield and Oxford. It was opened in 1884 but ceased to operate in 1930. The Waimakariri Gorge Bridge has a splendid setting where the river debouches from rocky cliffs onto the plain and today this spot is much favoured for picnicking and fishing.

It is understandable that a civil engineer should adopt concrete as the building material for his own residence. E.G. Wright went part of the way at Windermere. Sometime in the 1870s a Mr Worthington, who was engineer to the Levels Road Board, had an attractive house built on the outskirts of Pleasant Point on today's State Highway 8. Of single-storey construction, it has a verandah on the front and the north side. The corrugated iron roof consists of twin gables facing south and abutting another at right angles. The walls of plain concrete are plastered on both faces and generally are in a sound state. In 1883 it passed to one named Maze and until recently has remained in this family connection.

Worthington no doubt gained from this experience with concrete for he soon had a larger house built a short distance away in Te Nga Wai Road. This building, set well back from the road, is of two storeys with 350-mm concrete walls plastered on both sides. Square in plan, it has dormers on each elevation with paired sashes set below the eaves. This gives the house an unusual and somewhat graceless appearance. The lower doors and windows have concrete frames detailed in the form of dressings and there is a raised band like a string course at first-floor level on the elevations. The roof is corrugated iron. A noteworthy feature of the interior is the use of pressed metal ceilings which have a more flowing pattern than usual, showing the influence of the Arts and Crafts Movement and William Morris. For the 40 years of this century when it was a boarding house it had a more spartan interior, but the present owners have made a start on bringing it back to its former state.

An unusual example of early concrete can be seen at Barrhill in Mid-Canterbury where several buildings in this material date from the 1870s as the core of a planned village. It was the dream of John Cathcart Wason, a Scot from Ayrshire who

Egmont Brewery, King Street, New Plymouth. Built in 1877 and later known as the Taranaki Brewery, this is a very early use of concrete in an urban industrial building.

had bought over 8,000 hectares of the Lendon run, naming his property Corwar after his home in Scotland. After erecting a twenty-roomed single-storey timber mansion, he had a small concrete lodge built at the entrance to the long drive. This building consists of four rooms with concrete dividing walls and neither hall nor porch. With its simple square plan surmounted by a pavilion roof covered in slates, this diminutive building has the appearance of a doll's house. Built about 1875 it has been donated in recent years to the district and furnished in period style.

A short distance to the east Wason built his village of Barrhill, laying it out with avenues of oaks, birches, sycamores and poplars and a market square in the centre. He was a great lover of trees and did much other planting on his run. The focal building is the Church of St John the Evangelist built in 1876 and opened for worship on 8 July 1877. It has a concrete floor and walls 300 mm thick, with the roof supported on kauri trusses. Its setting is delightful and the church, still in use, is much admired.

Nearby is the school, also built of concrete with the same wall thickness. A large fireplace on the east wall heats air in pipes behind the wall linings, which are vented for warm air distribution. Today the building is used as a craft shop. Alongside is the school house, likewise of concrete and built in 1878 for the school opening. The other houses and buildings have been removed over the years as the village declined with the changing patterns of transport and work opportunities.

A remarkably early use of concrete in an urban industrial building is part of the former Taranaki Brewery in Queen Street, New Plymouth. As early as 1864 the Egmont Brewery had been established in nearby King Street by Henderson and Paul. A wooden building, it was replaced in 1877 on a new site by J. Paul and Coy (Henderson had moved to Wanganui). The concrete brew tower, which is sited alongside the Mangaotuku Stream, had plant giving a capacity of fifteen hogsheads and provided employment for nine men. Although modern equipment had been installed in a newer building, the original concrete structure still had some of its plant in the mid-1980s. The concrete building almost abutting it is of later construction and was used as a cool store.

After building his own house in concrete W.H. Clayton, Colonial Architect, had strong feelings that this was an

BLENHEIM POST OFFICE, MARKET SQUARE. Built in 1877 to the design of W.H. Clayton, Colonial Architect, this early concrete public building was demolished in the 1960s.

important new building material, for he used it in his design for the two-storey Blenheim Post Office of 1877. Unfortunately this was demolished as an earthquake risk in the 1960s. If it had been retained a few more years it would have been practicable to strengthen it using new techniques. The old post office certainly produced better townscape than its successor, for its scale, disposition and setting made it a key element in Market Square.

While preparing his design for the large Government Building in Lambton Quay, Wellington, Clayton considered concrete as a possible material and also stone, but shortage of money dictated that timber be used as the cheapest option. It is interesting to reflect that the present well-designed and executed building might have been built in concrete — tenders called in November 1873 gave the lowest price for concrete as £40,000 compared with £29,975 for timber. Clearly the size of this building, and the considerable formwork necessary, would have been a deterrent at that time when few contractors had any experience of concrete.

Clayton prepared drawings in 1876 for a prison at New Plymouth which is still in use. A concrete building, it was able to house 50 male and 10 female inmates. The prison walls surrounding it are stone. Another early penal institution was the Timaru jail, built in 1879 of concrete. It was demolished in 1914.

One of the interesting concrete structures of the 1870s, and an early industrial example, is the flour mill store at Ngaruawahia. It is on the true right bank of the Waikato River, just below the 'Meeting of the Waters' at the confluence with the Waipa River. There had been an earlier flour mill on the site, brought from Lake Waikare in 1871, but this timber building was destroyed by fire. It was replaced by a three-storey brick structure, and nearer the river a two-storey concrete store was erected to hold the finished flour and other products. Designed by the architect Thomas H. White, who was farming at Taupiri, it dates from about 1878 and measures 27.4 m by 7.6 m. The ground floor walls are 457 mm thick and above this reduce to 304 mm. Rather a plain building, as its simple function would suggest, it has considerable claim to fame in having some elementary reinforcement in its construction. White claimed that it was the first reinforced concrete building in the

GRAIN STORE, ST ANDREWS, SOUTH CANTERBURY.
The New Zealand and Australian Land Company built this store in 1878 to hold grain from their Pareora Estate. Sited alongside the railway, it is now an enigmatic ruin.

Southern Hemisphere, for he used single-strand barbed wire. In terms of technological and industrial heritage interest, it has been given a high rating by the New Zealand Historic Places Trust.

A surprisingly early design for a concrete wharf was that prepared by the engineer John McGregor for the Clive Grange Estate and Railway Company Ltd. This newly floated company of prominent Hawke's Bay sheepfarmers intended to build, in 1879, a major port in the lee of Black Reef near the cliffs of Cape Kidnappers. There was to be a concrete wharf of some 760 m in length with a railway and cranes. McGregor was the engineer for the Oamaru breakwater of 1872 and later became harbour engineer to the Auckland Harbour Board. The Clive Grange port never eventuated because the estimated cost of £60,000, in addition to £66,000 for the purchase of the Clive Grange Estate, proved to be excessive at a time of financial depression in the colony.

In North Canterbury a concrete cookshop was built sometime in the 1870s at Okuku Pass station, once part of the Whiterock run. It was of lime concrete with lime plaster on the exterior ruled off into 600 mm by 250 mm blocks simulating

FLOUR MILL STORE, NGARUAWAHIA.
Designed by architect T.H. White, this building dates from *c.* 1878 and is noteworthy for the early use of some elementary reinforcing in the concrete.

COOKSHOP, OKUKU PASS STATION, NORTH CANTERBURY.
Built in the 1870s in a remote area, this farm building used lime concrete, with the plaster finish ruled off to give a hint of stone work. There are two similar cookshops in the district.

stone. There are two similar cookshops built by the same person, one at The Brothers and the other at Journey's End and all are extant. They were used as musterers' huts.

Having discussed several examples of concrete farm buildings in this chapter, it is of interest to note that the earliest English counterpart still standing is a large 'barn' built in 1869 by Robert Campbell on his Buscot property in Oxfordshire. He was a very progressive and innovative farmer who had made his money in Australia where he farmed at Duntroon near present-day Canberra. Robert Campbell invested in new buildings and in schemes for intensive breeding, especially Indian cattle. This concrete byre was to house these animals. The roof covering and ends have been modified but otherwise the building is reasonably intact. The board marks from the shuttering are clearly visible. Campbell's son, Robert, was sent to New Zealand as a very young man to buy land on his father's behalf and such well-known properties as Galloway, Benmore, Station Peak, Otekaieke and Burwood were some of them. However, Robert junior did not introduce any concrete structures to his properties, which were under the control of managers. The local schists of Otago were already in use as were timber and corrugated iron.

The New Zealand and Australian Land Company had its origins in 1859 but really became effective in 1877 after amalgamation with the Canterbury and Otago Association Ltd. The driving force had been William Soltau Davidson, who later directed the company from the Glasgow head office, while, in New Zealand, affairs were managed by the dynamic Thomas Brydone. With sixteen properties there was considerable cropping undertaken and in order to hold grain from the Pareora estate a substantial store was built in 1878 near the railway line at St Andrews. Today it is a concrete ruin in a paddock. The walls rise about 2 m above the concrete floor. This appears to have been their original height and they consist of 200-mm-thick no-fines concrete with cement render on both faces. On the long elevations the company's initials and the date of erection can still be seen inscribed in the plaster.

On 30 July 1879 the foundation stone was laid for St Mary's Anglican Church in Southbrook, now part of Rangiora. It was consecrated in the following year. Designed by C.G. and C.J. Chapman it is a concrete building measuring 10.9 m by 5.5 m in a modified Gothic Revival style with narrow lancet windows. The concrete buttresses are probably for traditional architectural effect rather than for stiffening an unreinforced concrete structure of no great height. The exterior walls are plastered and the roof is of corrugated galvanised iron. Although a small church of simple design, it is an attractive presence in the street.

CORWAR LODGE, NEAR BARRHILL.
J.C. Wason, who had acquired part of the Lendon run on the south side of the Rakaia River, had this delightful doll's-house-like lodge built at the entrance to his property. His twenty-room house was built in timber.

Church of St John the Evangelist, Barrhill.
A focal point in the village conceived by J.C. Wason, the church opened in July 1877. Set in attractive grounds with mature trees it is a delight, both inside and out.

St Mary's Church, Southbrook.
This small Anglican church was begun in mid-1879 and was designed by C.G. and C.J. Chapman in a simplified Gothic Revival style. The buttresses are probably an ecclesiastic design element rather than a structural necessity.

SCHOOL, BARRHILL.
Opened in 1878, with the concrete schoolhouse alongside, the school was an integral part of the village. At the time of writing it was being used as a craft shop.

Chapter Four
The 1880s

Possibly dating from the seventies is a two-level concrete cookshop at Castlerock station near Lumsden in Southland. Pages from the *Illustrated London News* line a wall and show the date of 1893, but these could have been added later. The building was certainly in existence by 1881. It is rectangular with gable ends and a corrugated iron roof. The ground floor had a living area, a central kitchen and a bunkroom, while the steep narrow wooden stair led to the head shepherd's room.

Exterior walls where the plaster has peeled show evidence of hoop iron (presumably from wool bales) laid horizontally about every 300 mm. Left unused and without glazing for a long period, the building has recently been put into sound condition and modified to provide comfortable farm-stay accommodation — a worthy example of conservation.

Castlerock station, originally known as The Elbow, had its origins when Run 181 was taken up in 1859 and first granted to J.P. Taylor. In January 1866 it was transferred to Mathew Holmes, with Thomas Barnhill as the manager. After two periods on other stations Barnhill bought the homestead block of Castlerock when it was sold in 1903.

James Shand, the progressive farmer at Abbotsford near Outram who used concrete for his 1870s farm buildings, also erected another block of such buildings at his Berkeley property near Henley. Designed in early 1881 by the same architects, Mason and Wales of Dunedin, these new buildings formed a steading in a hollow square. They consisted of cartshed and toolsheds, cattle stalls, stables, shearing shed, and a killing shed. It is interesting that cattle were housed in stalls during the winter — a most unusual practice in New Zealand, although Alexander Mathieson at Centre Road near Dunedin had an even larger cattle byre. The swampy ground, which was drained to allow the buildings to be erected, was subject to heavy flooding so that over the years the foundations failed. Following the disastrous flood of 1981 it was decided to demolish the buildings. No doubt the lack of reinforcement and possibly inadequate footings for the conditions led to a lack of resistance to the stresses induced by very poor ground conditions.

After G.H. Moore had built his grand home at Glenmark station, he set about having a very substantial stable erected in 1881, using concrete throughout except for the roof. The architect, S.C. Farr, having already displayed his interest in concrete in part of the homestead, and being aware of its resistance to fire, chose concrete as the most suitable material for the stables. One of the largest buildings of its kind in the colony, it measures some 76 m in length and has 17-m-long wings at each end. A loft runs full length, as well as over the wings, for feed storage. The entire front elevation is arcaded with segmental arches to give a fine sense of articulation and lightness to the building, for all its size. The planning provided for a granary in the first wing with large sliding doors for access. Next to this was the smithy, adjoined by the groom's room in the corner of the main portion of the building. There was an area of nine bays for implements and then eight more bays with

STABLES, GLENMARK, NORTH CANTERBURY.
Designed by S.C. Farr, the 76-m-long stables building is one of the largest of its kind in the country. Completed in 1881, it contained granary, blacksmith's shop, stabling, carriage shed and a huge loft. For a time there was a shearing board in place of some stabling.

COOKSHOP, CASTLEROCK STATION, NEAR LUMSDEN.
This building may date from the 1870s. It was certainly being used by 1881 to feed and accommodate station hands, the head shepherd's room being upstairs. After a long period of disuse, it has been put into good order and provides comfortable, self-contained accommodation for travellers.

GATEWAY, GLENMARK, NORTH CANTERBURY.
The extensive use of concrete at Glenmark is seen in its impressive entrance.

loose boxes for the working horses. At some stage part of this area was adapted for a shearing board but recently this has been removed. Beyond this and extending to the end of the structure were six stalls, presumably for hacks but now used as a workshop. Projecting at right angles is a double-gabled wing with a carriage shed, water tank and an open arcaded area with troughs for watering horses. Steps lead from here to the loft.

It is clear that considerable thought was given to the efficient planning of this building. The materials of construction, plastered concrete and unpainted corrugated galvanised iron roof, have proved to be sound choices, for it is only after a century of use that some major maintenance has become necessary. Farr was without doubt a fairly pragmatic architect and his clients were well satisfied with his work, which he supervised with diligence. In this instance he not only produced a most handsome farm structure but one which must take its place as an outstanding early application of a new building material. The approach to Glenmark is from a delightful wooden Gothic lodge and follows a drive skirted by an attractive fence of concrete supporting iron standards. Gate posts are also of ornamental concrete. The Farr-designed bridge has concrete piers and abutments. Beneath it is a weir for the small hydro-electric generator housed in a shed on the stream bank alongside, an indication of Moore's progressive attitude towards new technology.

An unusual building is the disused chicory kiln at Inchclutha in South Otago, built in 1881. It stands isolated in a paddock close to the Clutha River and is a landmark clearly visible from Balclutha. Now rather derelict this plain concrete structure has a slightly sunken ground floor with a narrow gallery — a low-ceilinged work space surrounding the three brick furnaces with iron doors. Above these are iron poles and horizontal bars with perforated iron plates to hold the chicory roots while drying took place. An upper level provides access to handle the dried chicory at the first stage before it continued on the lower grid. A second drying floor, built of timber and asbestos cement, was added in 1929.

CHICORY KILN, INCHCLUTHA, SOUTH OTAGO.
Built in 1881 for Gregg & Coy, this interesting structure was designed by Mason and Wales of Dunedin. There are three brick furnaces.

Designed by Mason and Wales, the pioneer architects of Dunedin, the kiln structure was built for Gregg and Coy. It shows a confident belief in the merits of mass concrete. The plasticity in form is well illustrated in the semi-arched ceilings above the ground floor, with their rib vault intersections at the corners. These have an iron bar cast in, thus avoiding a sharp edge. Chicory was grown as a root crop which, when harvested, was diced, dried, and then in a separate factory ground and roasted as an additive to coffee.

One of the more uncommon concrete structures of this decade was the three-storey Firth's Tower built in 1882 by Josiah Clifton Firth on his property near Matamata. Today it is surrounded by various old buildings brought on to the site to form an historic village museum — a complex which unfortunately causes Firth's Tower to lose the rural setting it had for almost a century. J.C. Firth was an early settler in the district. He had very extensive areas of land, which he farmed intensively using the latest equipment and large-scale drainage to gain high production of crops. His estate soon came to be regarded as a model farm and aroused widespread interest. At his own very considerable expense he had the Waihou River cleared of snags so that river steamers could navigate it and transport his produce. Although he enjoyed good relations with Maori he

Opposite: WATER TOWER, ADDINGTON, CHRISTCHURCH. Built in 1883 for the New Zealand Railways Workshops (recently demolished), this is a pioneer structure using reinforcement well before it became generally used.

FIRTH'S TOWER, MATAMATA.
Once a landmark in J.C. Firth's extensive agricultural holdings, the tower was built in 1882 to T.H. White's design.

had his tower constructed near the estate homestead, possibly as a safeguard against possible further disturbances. It is highly probable that Firth had a romantic yearning to have a tower at his residence, for his Auckland home incorporated a four-storey tower as we have already noted.

The 18-m Waikato tower, an impressive four-square structure of concrete, cost £1,600 — for those days a considerable sum for a building that had little real use. The top storey oversails slightly, being supported on corbels — a carryover from brick and stone detailing — and has a pavilion or pyramidal roof of corrugated iron, surmounted by a small timber lookout. Doors and windows have prominent label moulds. Floors are of timber. The tower was designed by Thomas H. White, the architect responsible for the Ngaruawahia flour mill store and an early exponent of concrete construction. Although Firth's Tower was built four years or more later than the store, there is no reliable record of it having any reinforcement.

Another remarkably early example of ferro-concrete (concrete reinforced with iron or steel) and said to be the first real use of reinforcing in New Zealand is the water tower built in 1883 at the former New Zealand Railways Workshops at Addington, Christchurch. It was designed by Peter Ellis, then Chief Draughtsman for the Railways Department; his brother was the foreman during its construction. The reinforcement took the form of several tonnes of scrap steel, but what sort of sections were used — such as rods, bars, plates or angles — is not known. These were placed at 300-mm intervals. The

PETTIGREW HOUSE, OPUNAKE.
In 1883 William Pettigrew, an early settler, built his single-storey house of lime concrete. Originally it had a concrete slab roof, but another storey was added in 1984.

tower sat on a layer of quicksand of considerable depth. Instead of excavating through an overlay of clay, the method was to remove the loam at the top and place the structure directly on this clay bed above the quicksand. Initially there was 230 mm of settlement, which was predicted by the designer. This 18-m tower, which was used until quite recently, was built by prisoners from the Addington Prison. It is of interest to note that the first reinforced concrete water tower to be constructed in Britain was in Bournemouth in 1900 using the Hennebique system and this, too, is still standing.

Although completely hidden from view, concrete was used in the old suspension bridge at Alexandra. The construction posed problems for the founding of the tall pier on the town side of the Clutha River. In mid-1879 the foundations were excavated by removing a thin layer of schist rock sitting on 0.6 m of soft clay resting on a conglomerate of decomposed rock, quartz and slate. On this latter base was poured a 1-m concrete pad. It was a difficult operation, requiring three gangs of workmen at a depth of 6 m below normal river level. A threatened strike and spring floods compounded the problems of the contractor, Jeremiah Drummey. When the concrete foundation was ready, the masonry pier was constructed of ashlar (dressed stone in courses), having three superimposed arches diminishing in width and increasing in height. The Alexandra Bridge was designed by Leslie Duncan Macgeorge, Engineer to the Vincent County Council, with some assistance from the well-known Dunedin civil engineer, Robert Hay. It was opened in June 1882 and was regarded as one of the most attractive bridges in New Zealand. When a replacement steel arch bridge was built in 1958, public interest prevented the demolition of the two piers, which remain today as monuments classified for preservation by the Historic Places Trust.

Some early buildings relied on concrete made with lime instead of Portland cement. The result was not as strong, being inclined to crumble eventually. In 1883 W. Pettigrew, who had settled in Opunake, built his family a house made of lime concrete. A single-storey dwelling, it was modelled on the family home at Pollokshields in Glasgow. It had, however, a concrete slab roof which was later covered by an iron clad pitched roof. In 1984 a new owner stripped the roof in order to add a large upper floor providing a self-contained residence. The original roof was revealed as mass concrete using large river-rounded stones up to 140 mm thick. Today the original form of the Pettigrew house is difficult to envisage but the structure is still basically sound.

Although John Wilson and Company had engaged in building concrete houses in Auckland in the 1880s, there was at least one citizen who went his own way. John Edward Taylor arrived in 1881 after having been a successful businessman in Bradford, Yorkshire, where he was born in 1847. Soon after coming to Auckland with his brother he bought land in the Mangere Bridge district and proceeded to farm his property, which he called Water Lea. Although very much the gentleman

WATER LEA, AMBURY ROAD, MANGERE BRIDGE, AUCKLAND.
Built by J.E. Taylor, who farmed on this land, and believed to date from 1883, Water Lea is complemented by a small concrete stable and a concrete barn.

farmer, he built, largely with his own hands, the handsome and substantial single-storey concrete house still standing today.

The date of construction is not certain but was most probably about 1883 and would have been soon after the farm was established. The house echoes the wooden colonial villa, having verandahs along part of the front, north side and rear. J.E. Taylor had acquired some engineering skills and possibly these induced him to use concrete, or he may have been unfamiliar with timber construction and distrusted its permanence. Certainly his house is well designed and constructed, with little evidence today of deterioration. There is still a half-hectare of land surrounding the house, with a garden and lawns containing mature trees. Two concrete outbuildings date from the 1880s — a small stable and a two-level barn with a barrel roof. The entrance to Water Lea is graced by fine wrought iron gates with large concrete posts and flanking walls.

John Taylor, a liberal in outlook, was a man of strong convictions with an intense interest in good working conditions and an eight-hour day for workers. He was prominent in the temperance movement and in public affairs. He advocated the use of motive power obtained from the tides and was a staunch believer in the idea of a canal linking the Manukau and Waitemata Harbours through the Tamaki River.

A far cry from the spacious Water Lea is the tiny four-roomed cottage built in 1883 at 1 Bankside Street off the top of Shortland Street in Auckland. A wooden cottage had been erected on the section sometime between 1854 and 1861 but was subsequently removed. The section was purchased by Dr Frederick Wright, who built the existing cottage in concrete. On 11 October 1883 he sold it to John Mulvihill, an Irish-born immigrant from the USA who had settled in Auckland with his wife Mary. During their occupancy it was home for five children and it remained in the family until 1924. After several ownerships the cottage was bought by the Auckland City Council in 1984.

The construction consists of concrete foundations and walls using Portland cement, which would have been imported from England as local Portland cement was not produced until the following year. The exterior walls are approximately 300 mm thick, whereas the interior partition walls are 150 mm. All wall surfaces are plastered using a lime mortar, except in the roof space, which shows the board texture of formwork. This consisted of 300-mm-wide boards. The timber-framed roof was of corrugated galvanised iron with the laps sealed in pitch. The floors are timber. The approach is direct from Bankside Street up a flight of steps to a full length verandah with concrete floor and wrought iron balustrades of ellipses and vertical rods. The rear also has a verandah, with the south end filled in to house a water closet.

1 BANKSIDE STREET, AUCKLAND.
Dating from 1883, this four-roomed cottage is now owned by the city council and retains its structural integrity.

The value of this cottage lies in its undisturbed structural condition and the fact that it has survived intact in the heart of commercial Auckland. There has been no alteration over the years to its form or materials of construction other than for essential maintenance.

The world's first concrete lighthouse dates from 1873 in Jersey in the Channel Islands and is known as Le Corbière lighthouse. The New Zealand equivalent was built in 1882 when Burgess Island in the Mokohinau Group was chosen for a light to serve shipping in the Hauraki Gulf. Because the local stone was found to be unsuitable, the decision was made to build the lighthouse in concrete. Rockbound Burgess Island is about 24 km northeast of Great Barrier Island so that the remoteness and steep site influenced John Blackett, Assistant Engineer-in-Chief of the Public Works Department (and also Marine Engineer of the Marine Department), in constructing purpose-made concrete blocks in Auckland. These cast blocks were transported to the island and winched to the summit for the building of the lighthouse. It came into operation on 18 June 1883 and there can be no doubt that it was a pioneer use of precast concrete elements in building construction in the colony.

John Blackett was a remarkable engineer who was responsible for a wide variety of engineering structures throughout the country. In a close partnership with the Marine Department's Nautical Adviser, Captain Robert Johnson, he produced twelve lighthouses between 1870 and 1880 and seven others after this decade. He later became engineer-in-chief of his department.

As mentioned in the previous chapter, an early all-stone graving dock built in 1872 at Port Chalmers is no longer visible, having been filled in for container storage. The Lyttelton graving dock, formally opened on 3 January 1883, also has stone construction. However, this rose on the sides for only eight altars, stepped to give the characteristic form of such docks. Above the stone, the material is poured concrete in much larger altars so there is a sharp contrast in scale. I do not know whether the supply of suitable stone became scarce or whether the more innovative material suddenly found favour. Certainly there is harmony between the two materials. The superb sculptural qualities of this dock are better appreciated when viewed from the more elevated Brittan Terrace overlooking it.

The engineer to the harbour board, Charles Napier Bell, was responsible for its design and execution in which he achieved a splendid amalgam of function with robust elegance. C.N. Bell (1836–1906), a Scot, came to New Zealand in 1871 as an engineer for Brogden Brothers, an English firm with several railway construction contracts in various parts of the country. As Lyttelton Harbour Board engineer he designed and built the patent slip and several wharves. After this he carried out many engineering investigations in New Zealand, especially for harbour works, and later in Australia. He was one of a breed of outstanding civil engineers of the period.

The second synagogue to be built in Auckland was that designed by Edward Bartley in 1884 and dedicated in the following year. The building used concrete made of hydraulic lime throughout and displays predominantly Romanesque features. It is of two storeys with a basement and is roofed in slates. The interior has galleries and richly ornamented plasterwork in the apsidal space for the Ark. The ceilings, barrel vaulted in the main area and pyramidal over the former women's gallery, are decorated in stencilled patterns of green, red, and pink on a cream ground. After being abandoned for many years it has recently been sensitively restored and converted to a bank with a well-proportioned small single-storey addition on the north side. Edward Bartley trained as a builder as well as an architect and he practised for many years in

Auckland. If he were alive today he would assuredly approve the fine handling of the strengthening, modifications and refurbishment.

Another concrete building of 1884 was Overton, a country house built for the Hon. Francis Arkwright and situated several kilometres from Marton. It was designed by Frederick de Jersey Clere, an English-trained architect practising in Wanganui with Alfred Atkins. Later he came to Wellington where he established a long-standing practice. Clere used the Elizabethan half-timbered style to produce a gracious homestead constructed of 200 mm by 200 mm heart totara framing with the infill of Portland cement concrete in the place of brick nogging or wattle and daub. A contemporary newspaper account stated that the architects had employed devices to ensure that the concrete would not fall out in the event of an earthquake and this has certainly proved to be the case. There was a reference to the fact that composite houses of concrete cost from 25 to 30 percent more because they were not yet in general use. Overton is a two-storey house with dark painted, exposed framing and a corrugated galvanised iron roof. The windows are leadlights. Francis Arkwright had taken up land here in 1882 when he came out with his brother-in-law and family. Francis had an illustrious great-great-grandfather in Sir Richard Arkwright, the British industrialist and inventor of the water frame in spinning. Francis was an accomplished French scholar, having translated the memoirs of the Duc de St Simon. He was called to be a member of the Legislative Council in 1895 and was also a strong supporter of Richard John Seddon. A few years ago Overton changed ownership out of the family but the integrity of this fine house is being respected.

Another impressive house of the same period is Green Hayes in Temuka, built in 1886 for John Turnbull Murray Hayhurst, son of John Hayhurst who was a well-known runholder. J.T.M. Hayhurst took over the Green Hayes farm from his father and became landlord of 30 adjoining tenant farmers. He also had a flour mill and substantial commercial properties in Temuka. The two-storey house of sixteen rooms, or mansion as early accounts describe it, was built of 450-mm-thick concrete and is notable for the splendid plaster work in the porch, hall and main rooms. An innovation at the time was the illumination by electricity generated by a turbine using water from a nearby creek. The house was graced by conservatories flanking both sides but these no longer exist. For many years it has been used by the Salvation Army as a children's home. There has been a substantial addition, and this, together with other changes such as external fire escapes, has diminished the splendour of the house.

The main hop growing area in New Zealand is the Nelson region from Riwaka to Waimea West and south to Sherry River. Today there are twenty concrete hop kilns in existence, the oldest (dating from *c.* 1885) being known as Bradley's after the original owner. It is at Dovedale and was unusual in having a shingle roof — a strange combination of traditional and new materials. Already in a rather derelict condition, this roof collapsed in a storm in 1989 and the disused kiln now has only the concrete walls as a shell. A more complete extant example is Ewer's Kiln, built about 1900 in Upper Moutere (see pages 93 and 94).

Concrete is an ideal material for constructing fortifications and it was used extensively to build military defences to guard the entrances to the main ports during the so-called 'Russian war scare' of the mid-1880s. During the time he was Governor of New Zealand (1882–89), Lt.-Gen. Sir William Jervois maintained an enthusiastic advocacy for his system of coastal defences as a continuation of the British policy of ensuring freedom of the seas for its merchant fleets. Regarded as Britain's foremost defence works strategist and military engineer, he advised on fortifications throughout Britain and its possessions. His plans for the defence of the harbours of Auckland, Wellington, Lyttelton and Dunedin were achieved with the help of several able engineers. In 1880 Lt.-Col. P.H. Scratchley, Royal Engineers, had recommended a scheme for these harbours.

In Auckland, the defence area in Devonport, known as Fort Cautley, has within it Fort Takapuna containing a structure of concrete with some brickwork erected in 1886–87. The building, set below ground level in an excavation, is approached by steps and has a narrow walk space round part of it. Inside is a barrack room and kitchen, with tunnels leading to ammunition storage areas and gun emplacements. The main block has concrete retaining walls about 600 mm thick although the other walls are brick. Floor and roof slabs are concrete as are the gallery (tunnel) linings, with the roof slabs reinforced by iron rails — an improvement on the brick arches used hitherto. Bounded by a crenellated concrete parapet, the roof originally had a covering of earth as camouflage. A scheme for renovation and restoration by the former Ministry of Works and Development is in abeyance. The name Fort Cautley comes from Major Henry Cautley, Royal Engineers, who had been assistant to Sir William Jervois in military defence proposals.

The key fortification in the Wellington Harbour network was Fort Ballance on Point Gordon, Miramar (Watts) Peninsula. It is still in existence although several gun emplacements have been imploded to prevent access. A barrack block built against and into a bank is mainly concrete with some plastered brickwork. Loopholes penetrate the main outer wall. The several gun emplacements and retaining walls are concrete, with some plastered brick. Two high walls have permanent shuttering of now rusted corrugated iron secured to iron rails set into the concrete. They had loopholes for rifle fire. Rails from the Manawatu Railway Company were used as reinforcement for the gallery roofs. Construction began in 1885 but was not finally completed

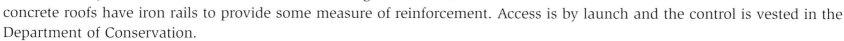

FORT BALLANCE, POINT GORDON, WELLINGTON. As the principal fort in the harbour, this fort has a barrack block, partly underground, and several gun emplacements. Construction dates from 1885 to 1893, and was under the control of the Public Works Department engineer, A.W.D. Bell.

until 1893. Although Captain (later Major) Edward Tudor Boddam came over from Australia to replace Major Cautley as Engineer for Defence, he was not a qualified engineer. Much of the work was carried out under the Resident Engineer for the Public Works Department, Arthur Wilbraham Dillon Bell. He was in charge of defence works in Dunedin and in 1888 he was responsible for all such work throughout New Zealand. Recently the Army has undertaken to restore the complex to a state suitable for public access — a commendable exercise ensuring the preservation of an important part of our military history in a dramatic setting.

Lyttelton had a fine fortification, Fort Jervois, on low-lying Ripapa Island. Today it is still complete and most impressive. The emplacements for the 6-inch and 8-inch Armstrong breech-loading disappearing guns and the galleries are concrete, but the barracks are of masonry, as are the extensive walls surrounding the fort. The concrete roofs have iron rails to provide some measure of reinforcement. Access is by launch and the control is vested in the Department of Conservation.

At Fort Taiaroa, guarding the entrance to Otago Harbour, the galleries are lined in stone but the gun emplacements are concrete. The Armstrong disappearing guns were installed in these and one has been maintained in good order.

Rural districts throughout New Zealand feature a surprisingly large number of concrete buildings erected during the nineteenth century. One such is the Appleby house, built by Alex Ross for his own use and located on a small farm at Te Awa, a few kilometres south of the Maramarua State Forest in the northern Waikato. Dating from 1886–87 this single-storey house has a rather porous concrete mix suggesting lime concrete. It is plastered on both faces and is lined out on the exterior to simulate masonry. Now festooned with creeper and no longer a family home, it does not present its best features. The original corrugated galvanised iron roof has been replaced by galvanised steel panel deck. However, the style of the house is remarkably undated and could seemingly have been built in the 1930s or later.

From 1880 John Wilson and Company of Warkworth began building houses in hydraulic lime cement. In the Grey Lynn suburb of Auckland several of these early concrete residences are still in use. Two are on adjoining sections. At 350 Richmond Road a substantial house was built for Richard and May Warnock, probably in 1886 soon after they purchased the land. The building is single storey at the street elevation but the sloping land allows two high storeys at the rear. From the front it is a bay villa but with slate roof and quoins formed in the cement plaster finish. There is further decoration in the bracketed eaves. The bay windows have round-headed top sashes whereas the others have segmental arches. There are full-length verandahs on both floors at the rear and today these show partial enclosures. Altogether it is a handsome house and no doubt the owners were proud of it, especially with its use of a relatively new building material. Family records state it was left vacant for twelve months after completion to allow the concrete walls to dry out thoroughly before occupation. One surmises this was a reflection on Auckland's humid climate. The concrete house next door is generally similar in size and design. However, it is surrounded by mature trees and rather difficult to see in its entirety.

A farmhouse known to have been built of concrete sometime in the 1880s is the Bowmar house in the Tara district, about 7 km north of Mangawhai in Northland. A rather modest house, it is L-shaped with attic bedrooms in both wings. The chimneys are brick and floors are of wood. Local legend has it that green manuka sticks were used to reinforce the concrete, but there is no evidence of this in a broken windowsill exposing the mix. However, this is not to say that the walls haven't any such reinforcement. The same legend claims that the concrete was made from lime produced from burnt shells. Certainly the broken portion does indicate lime concrete, which was not too uncommon at that time. The first owner was Peter James, a Cornish stonemason who built a stone dairy in a paddock behind the house. Later the property passed to a Joseph Bowmar or possibly one of his many sons. After a period of neglect and abandonment the house is now occupied and well tended.

Concrete was used quite extensively for foundations and floors, especially where industrial processes demanded either very strong load-bearing surfaces or a measure of waterproofing for washing down. The Thames School of Mines, which began in 1886, also operated the Experimental Metallurgical Works. This consisted of a large space where a concrete floor was formed to accommodate a stamper battery, berdans (for crushing ore) and channels with sumps for the potassium cyanide process of gold extraction. The floor still survives with modifications over the years as the plant layout was changed. This space, today known as the Battery (Stamp Battery) Room, dates from 1887. Concrete was also used on the floor of the Assay Room and as a base for the Pelton wheel and generator in the Electrical Generating Room.

When Hugh McLean, a coachman from Ardross in Ross and Cromarty, Scotland, arrived in New Zealand in 1862, he no

BOWMAR HOUSE, TARA, NORTHLAND.
Dating from the 1880s the farmhouse is still in use and in good condition. There is no evidence to support a claim that green manuka sticks were used as reinforcement.

doubt had the ambition to buy his own farm. After 25 years he was able to acquire two properties just north of Amberley, calling them Glasnevin and Ardross. At the latter he built a two-storey concrete house in 1887. A partly demolished partition wall, about 200 mm thick, reveals 150-mm stones used as 'plums' in the matrix. At some later date he added a two-storey portion in timber with considerably higher studs and this was plastered subsequently to give a more homogeneous appearance to the exterior of the house. It is quite likely that S.C. Farr, the Christchurch architect who designed the concrete structures at Glenmark station not far away, was responsible for the concrete house. He was known to Hugh McLean through alterations and additions to his Crown Hotel in Amberley. McLean's son, Hugh Henry, took over Ardross farm, which had some 1,000 sheep by the turn of the century. The house is still in use.

The first all-concrete graving dock to be constructed in the colony was the Calliope Dock at Devonport. Built for the Auckland Harbour Board, the contract was let in November 1884 with the opening taking place on 16 February 1888. The designer was William Errington, Dock Engineer to the board, who had earlier been the engineer for the design of the Western Springs Pumping Station. The Calliope Dock became a most important facility, particularly for the Devonport Naval Base. It was capable of handling two large warships simultaneously. The original length was 160 m with a width of 25 m and a depth of 10 m. Some 25,000 barrels of cement were used in the construction. Since its opening the dock has twice been extended, being 185 m long at the present time. In 1987 the harbour board sold the dock to the Royal New Zealand Navy.

CALLIOPE DOCK.
A graving dock built for the Auckland Harbour Board, Calliope Dock was opened in 1888 and has been extended twice. It has been owned by the Royal New Zealand Navy since 1987. *(Photo: Royal New Zealand Navy Museum, Devonport.)*

GRAVING DOCK, LYTTELTON.
Designed by C.N. Bell, Engineer to the Lyttelton Harbour Board, the graving dock was opened in January 1883. The lower steps (altars) are stone, with concrete above. The dock is still in use.

FORT TAKAPUNA, FORT CAUTLEY, DEVONPORT.
Built in 1886–87 in reaction to the perceived threat of Russian designs on the British Empire's merchant fleets, this is one of several forts in the four main ports. It consists of the partly brick barrack block and galleries to the gun emplacements.

Opposite: SYNAGOGUE, PRINCES STREET, AUCKLAND.
This former synagogue, built of hydraulic lime concrete, was designed by Edward Bartley and completed in 1885. Vacant for many years, it was sensitively adapted for use as a branch of the National Bank. The interior has splendid ceilings.

OVERTON, NEAR MARTON.
The concrete infill of this Elizabethan-styled farmhouse, designed by F.de J. Clere for the Hon. Francis Arkwright, is secured to totara framing.

Opposite: GREEN HAYES, TEMUKA.
Built for J.T.M. Hayhurst as a large farmhouse in 1886, Green Hayes was originally flanked by conservatories. Now a Salvation Army children's home, it has been altered and extended.

350 Richmond Road, Grey Lynn, Auckland.
Built *c.* 1886, the house used hydraulic lime cement and is one of several erected in the area about this time.

Opposite: Ardross, near Amberley.
The farmhouse was built in 1887 by Hugh McLean, an immigrant from Scotland who arrived in New Zealand in 1862. The front of the house is a later addition in timber, plastered to match the concrete.

Chapter Five
The origins of cement manufacture in New Zealand

The 'father' of cement manufacture in the colony was Nathaniel Wilson. With his parents he emigrated from Scotland, reaching Auckland on 10 October 1842 on the *Jane Gifford*, one of the first two immigrant ships to that settlement. Born in Glasgow in 1836, he was seven years old when the family arrived in New Zealand. His father, William, was a blacksmith who later moved with his family to Kawau Island to work at the copper mine. After its closure they returned to Auckland where Nathaniel was apprenticed to the bootmaking trade. When he had completed his time he went to Victoria, Australia, to try his luck in the Ballarat gold rush. Not having been successful he came back to New Zealand, and as the family had moved to Warkworth Nathaniel set up there as a shoemaker. His father established himself at his own trade. In 1863 Nathaniel married Florence Snell and they lived in a small cottage of only two rooms and a workshop. With a family of ten children, they needed to enlarge their home over the years. In 1897 after their children had grown up and left the parental home Nathaniel built a substantial two-storey house in a form of concrete described later.

Some years after working at his shoemaking business, his health deteriorated and he was advised to seek a more open air life. Having 40.5 hectares of land he decided on farming and in 1866, a year later, he built a kiln for burning limestone found on his property. John Southgate already had a kiln further upstream on the Mahurangi River, nearer Warkworth. Soon Nathaniel had two kilns of 2.7 m diameter producing 'Roche' lime, which was slaked with water when about to be used and mixed with aggregate to form a concrete. When 'Roche' lime was ground by edge-runner mills with a proportion of scoria ash it was used for brick and plaster mortars. Having two kilns within three years, he was able to use one for drying hops which he grew nearby.

By 1878 the lime business had progressed to a stage that warranted setting up a family company. With his brothers James and John he formed John Wilson and Company, with these two moving to Auckland to establish an office, store and yard. In 1880, to publicise the merits of hydraulic lime, the company set about building concrete houses in various parts of Auckland, as already mentioned. Their product was used in the Synagogue in Princes Street and Firth's Flour Mill in Lower Queen Street as well as in other buildings. However, the practice proved disastrous financially and the company was almost bankrupt. By concentrating on manufacture only, John Wilson and Company eventually traded its way back to a reasonable position.

By 1883 the original two kilns had been replaced by as many as eighteen smaller kilns of 1.8 m diameter by 5 m high operating on coke obtained from the Auckland Gas Company. About this time a new steam boiler and 112 kW engine were installed, requiring a 15.24 m chimney. It was in this year that Nathaniel's friend, J.A. Pond who was Government Analyst in Auckland, told him of a book entitled *Science and the Art of the Manufacture of Portland Cement* by Henry Reid CE and suggested he try making some Portland cement. Nathaniel read this work and then began experimenting, but it was only after prolonged trials that he eventually produced a reliable product. It was the first Portland cement made in New Zealand and also in the Southern Hemisphere. There was, however, a strong body of opposition to the Wilson cement by many specifying authorities who preferred to rely on the imported brands. This lack of faith in New Zealand manufacturers was quite widespread in the Victorian period — much of it was a form of snobbery. It did considerable disservice to newly emergent industries and the opportunities they offered for work, especially in skilled occupations.

Nathaniel Wilson first burnt the Mahurangi limestone, ground it to a fine powder and then made it into bricks. When dried, these were burned to fusing point with coke. The hard, strong residue or clinker that resulted was then ground to a fine powder ready for marketing. Before the cement left the works Wilson carried out tests to ensure it complied with English specifications. He had built two large additional kilns measuring 9.1 m by 3 m diameter but then he experienced acute difficulty in achieving quality from the limestone available. After considerable experimenting, which used up capital with no corresponding output, he reluctantly admitted defeat and asked his old friend, J.A. Pond, to advise him. Pond was able to find a satisfactory solution for improving the limestone and so production was resumed and reached the high standard achieved earlier.

In 1894, with two new boilers being installed, a new chimney was required. This was built in hydraulic lime concrete to a height of 30.48 m with the top 1.5 m completed in Portland cement concrete. There was reinforcing every 0.76 m with bands of 380 mm by 127 mm iron but no vertical reinforcement. Another kiln was added at this time, being 15.24 m high by 6 m outside diameter and 3.6 m on the inside. It was built of hydraulic lime and shingle concrete, and reinforced by similar iron bands every 0.6 m in height. The cover of concrete was 150 mm on the exterior. Later there were three such kilns.

By 1897 Wilsons' production was some 1,524 tonnes per annum and in 1902 it had reached 7,620 tonnes. About 1898 or 1899 the company sent W.J. Wilson, its engineer and Nathaniel's eldest son, to the United States to study the latest developments in cement manufacture. A friend of Nathaniel was Norman Horsley, the company's agent in Christchurch, and he had recently visited the United States, returning with very favourable impressions of the American methods. W.J. Wilson's visit resulted in the company installing two 18.3 m by 1.8 m rotary kilns and a 6.7 m by 1.5 m tube mill. There was also a ball mill of 2.1 m diameter by 1.37 m. The first successful operation of a rotary kiln was in the United States, a patent having been taken out in 1896 by Hurry and Seaman. The very first patent was in 1877, in England, by Thomas Russell Crampton.

In 1903 Wilsons floated a public company and also rebuilt their works with some larger structures — one measuring 18.3 m long and 12.1 m high was built of 228 mm unreinforced concrete. The mix was eight of coarse aggregate to one of cement. Today these roofless structures form picturesque and impressive ruins. In 1907 a new rotary kiln, 30.4 m in length and 2.4 m diameter, was installed.

Wilsons had a wharf alongside their works on the Mahurangi River, enabling the cement to be shipped to Auckland. At first it was packed in barrels which required the services of a cooper in full-time employment. At some stage a changeover was made to cotton bags, there being six of these to the ton. Another change was made to jute bags; and in 1948 the multiwall paper bag came into use.

Several other companies involved in Portland cement production had varying degrees of success. At Ferntown near Collingwood some plant, including a crusher and kilns, was erected *c.* 1883, but for some reason no manufacture ensued and the works were sold. In 1906 at Motupipi, also in Golden Bay, cement was being produced. About this time limestone from the Tata Islands in Golden Bay was being shipped to Picton for manufacture into cement. A more successful venture was the Golden Bay Cement Works Limited which began operations in 1908 at Tarakohe. In 1919 the Golden Bay Cement Company Limited took over and continued its operations at the same site, with the works being adjacent to their own wharf for shipping cement to the North Island.

The first South Island cement works to actually produce Portland cement was that of James Macdonald, who began operations in 1887 at Pelichet Bay, Dunedin. In the following year the Milburn Lime and Cement Company Limited was established and it took over his equipment in 1889. This company continued production at Pelichet Bay until 1929 when a move was made to Burnside on the south side of Dunedin. In 1897 they became manufacturers of silica Portland cement, which was claimed to be superior to the standard cement. Milburn Lime and Cement had acquired the rights from the International Cement Company of Denmark and conducted successful tests under the control of F.W. Petre, architect, James Hislop, architect and Robert Hay, engineer, all of whom were from Dunedin and were well known and highly regarded in their professions. About 1908 the company leased from the government the lime works and deposits at Makaraeo near Dunback and today they operate large workings there for the limestone.

On Limestone Island in Whangarei Harbour cement works were built in 1895 by Rutherfurd and Coy. There was a deep-water wharf alongside for bringing in coal and shipping out the cement. In 1896 Alan Hall founded the New Zealand Portland Cement Company Limited, which took over the Rutherfurd company and continued production until 1918. The business was then transferred to Portland on the south side of the harbour.

In 1900 John Wilson and Company bought a small cement works with two kilns at Te Kuiti. However, the older company was attracted by the new works at Portland for they had the advantage of coal supplied from Kamo and Hikurangi as well as deep water for shipping out the cement to Auckland. In 1912 the Dominion Portland Cement Company was established with works at Portland by W.J. Wilson, who had retired from Wilson's Portland Cement Company Ltd (formerly John Wilson and Company Ltd) as engineer. He became works manager and engineer and with his partner, George Winstone Jr, visited the United States, England and the Continent to secure the most up-to-date machinery. They selected two rotary kilns 48.76 m in length by 3 m diameter as well as other equipment. The company commenced operations in 1915 or 1916.

In 1918 Wilson's and the New Zealand Portland Cement Company on Limestone Island decided to amalgamate and buy out the Dominion Portland Cement Company, as the Portland site had so much in its favour for future development. The new firm was known as Wilson's (NZ) Portland Cement Limited. Operations ceased at Limestone Island, while at Warkworth hydraulic lime was manufactured with intermittent production of Portland cement. In 1929 work finally ceased and the abandoned cement works, which became derelict over the years, are now a picturesque ruin open to the public.

Chapter Six
The 1890s

For some unaccountable reason there do not appear to be many surviving examples of concrete structures from the last decade of the nineteenth century.

About 1890 a small wooden cottage of the 1860s at Te Kauwhata in the Waikato had an addition built of concrete. It is single storey with a verandah on two adjoining sides. Today it houses the collection of the local museum society.

An unusual example of the increasingly varied use of concrete by this time is a set of races at the woolshed on Whiterock station in Canterbury. Built in 1891 the shed had an internal shearing board with chutes to the ground leading to the concrete races on one side. The reason for using concrete is not recorded. The walls, slightly splayed with a rounded top, are no longer used and have a decided tilt which appears to have been intentional. Tapering concrete piles support the woolshed to give a good clearance above ground level, foreshadowing the substitution of concrete for traditional piles of totara and silver pine. There was also a concrete dip.

We have already noted the early adoption of mass concrete buildings in the Kaikoura district during the 1870s. George Frederick Bullen, who lived at The Elms homestead at Kahutara, gave money for a new hall for the Presbyterian Parish of St Paul in Kaikoura. Built in 1892 it is a substantial structure alongside the older wooden manse and it has a cement render on both exterior and interior faces. It is still in use and appears to be in sound condition. It was an appropriate venue for an illustrated talk I gave in 1987 on the development of concrete in New Zealand. A plaque in the hall interior commemorates the donor.

Among the more impressive remains of a large stamp battery for gold mining are the foundations of the Crown Battery at Karangahake. Sited on the hillside of the gorge, the ruins today are devoid of superstructure and consist of massive concrete ramparts with some impressive stone walling on a higher terrace. Because the normal form of a battery was a series of connected structures 'cascading' down a steep slope to facilitate operations by gravity feeding of the ores, the foundations are terraced. Large stamp batteries required solid and deep foundations to support adequately the continuing pounding of the stamps onto the ore in the mortar boxes.

The Crown Battery was selected by the Cassell Gold Extraction Company as the site for the first field test in the world of the revolutionary MacArthur–Forrest process of potassium cyanide treatment. This greatly improved the recovery of gold and silver from the crushed ores. The test took place in July 1889, and in 1892–93 a new battery was erected. By 1898 this had been extended to provide crushing power from 60 stampers — certainly a very large crushing battery, but not directly comparable with the mighty 200 stamp operation of the Victoria battery erected a few kilometres east at Waikino by the

ST PAUL'S HALL, KAIKOURA.
The cost was met by G.F. Bullen of Kahutara and the hall was built in 1892 for the Presbyterian parish.

Waihi Gold Mining Company. Today the remains of the Crown Mine Battery are preserved as a goldfield industrial monument interpreted and managed by the Department of Conservation. The site is an optional part of a walkway.

One of the leading exponents of concrete was Dunedin architect and engineer, F.W. Petre. In 1892 Sacred Heart Church in North East Valley, Dunedin, was opened for worship. Here Petre adhered to a neo-Gothic design, having a timber-framed roof to his concrete building which is plastered on both faces. Simple in concept, it is graced by a small belfry above the entrance and is attractively sited in a corner section with a garden setting.

In 1894 Petre designed the splendid St Patrick's Basilica in Oamaru. The basilica form derives from early Roman times when it was used as a hall of justice. It was adopted by the early Christians for their worship in Rome and parts of the Roman Empire. The basilica design featured columns, with a clerestory above the nave, side aisles, a flat, coffered ceiling to the nave, and an apse.

The Oamaru church, sited in Reed Street, is slightly elevated and is approached by a flight of steps. Construction consists of concrete core walls faced on both sides by the renowned Oamaru stone, a high quality limestone. The exterior has weathered to a pleasing grey in a

SACRED HEART CHURCH, NORTH EAST VALLEY, DUNEDIN.
A neo-Gothic design by F.W. Petre was used for this church, which opened for worship in 1892. It is a simple but attractive building happily sited on a generous corner section.

ST PATRICK'S BASILICA, OAMARU.
This splendid church, designed by F.W. Petre, is built of concrete with Oamaru stone on both faces. It was opened in 1894 for worship. The ambience of the interior, with its fine coffered ceiling and warm tones of creamy limestone, is an enriching experience. The exterior follows the basilican form with entrance portico. An elegant dome over the sanctuary and the twin cupolas on open towers add to the splendour of the design.

composition that includes entrance portico, small rotundas capped by cupolas, and the grand but refined dome over the sanctuary. Inside, one is immediately impressed by the warm light reflected from the attractive pale cream limestone. The clerestory, coffered pressed zinc ceiling, and interior of the dome are all elements that engender harmony and ambience. Opening took place on 23 November 1894 but the building was not complete. In 1903 the portico and cupolas were added, and it was 1918 before all work was finished.

The traditional rabbit fence in New Zealand has been constructed over many kilometres of inland country to protect farming areas from the marauding pest. Although mostly in the South Island, there was one long stretch of fence erected in the 1890s by the Hawke's Bay Rabbit Board from the southern end of the province at the coast to the Manawatu Gorge. Such fences were of wire mesh usually taken into the ground to prevent burrowing by the rabbits. The seaward portion at Pipi Bank station south of Herbertville had a concrete wall below the high tide mark — another example of the many and varied uses of concrete in the last century.

The construction of the Nelson railway to the West Coast was a long drawn out affair, as indeed were many early railways because of the nature of the country and the shortage of money. In 1893 the Kohatu Tunnel No. 1 was completed, being 1.35 km long and lined in concrete blocks. This railway only reached Kawatiri and eventually it was closed in 1955.

About 1893 Alexander Campbell, who had taken up his Auchmore property on the Taieri in the 1860s, built himself a two-storey L-shaped house of mass concrete. The 250-mm thick walls are plastered on both faces. Originally a bull-nosed verandah ran full length along the front of the house but this was altered when part was removed to form a porch with a lean-to roof. More recently the living room has been extended into this area and these alterations to the exterior wall revealed

AUCHMORE, TAIERI PLAINS.
Alexander Campbell took up this property near Outram in the early 1860s and for a time lived in a concrete stable. About 1893 he built the L-shaped house, which has 250-mm-thick walls.

red ochre that had been mixed into the concrete for decoration while it was still plastic. This house is in sound order considering the lack of reinforcement.

At Waikino, on the true left bank of the Ohinemuri River, the Waihi Gold Mining Company had begun building the huge Victoria Battery in 1897. This was sited on relatively level ground and had massive concrete foundations which are all that remain of the building today. A bridge was built in that year to give access and its foundations were concrete. In 1898 a huge vat room was built and this held a row of shallow concrete sand-leaching tanks, each 50.24 m square. They retained sand which had some 'slime' consisting of pulverised rock and potassium cyanide solution for a treatment lasting thirteen days. Nearby are the concrete foundations for the B & M tanks built about 1910 and described later.

Although not a concrete building in the real sense, Nathaniel Wilson's home on the outskirts of Warkworth is worthy of mention. In 1897, in his later middle age, he built a large two-storey house. Riverina, as he called it, was a most unusual building in its construction. It incorporates fired clay, reputed to be a traditional

EWER'S HOP KILN, UPPER MOUTERE.
This is the second-oldest of twenty concrete kilns in the district, and was built about the turn of the century.

Cornish method. Clay was dug from a deep pit on the site and fired in it, using puriri logs, over a period of four weeks. The still hot clay was then rammed into formwork for the exterior walls of 225 mm thickness, reduced to 130 mm for the interior walls. This burnt clay resembling crumbly brick was plastered on both faces. The house has ten rooms and a large outbuilding at the rear is built of the same materials.

Recently Riverina has been restored. When purchased in 1969 by the present owner the walls were cracking and it became necessary to use 25-mm steel tie rods to restrain the movement.

By the end of the nineteenth century concrete bridge piers were commonplace. An example in a railway viaduct can be seen in George Troup's design for the Kopua Viaduct over the Manawatu River, built in 1897 about 3 km north of Ormondville in Hawke's Bay. The two shore bays of plate girders are supported on concrete piers, whereas the taller midstream piers are of steel lattice construction.

The visual interest of suspension bridges lies in the clear span and the catenary of the cables from which the hangers support the deck with its stiffening truss. However, an integral part of the structure are the towers, which take the weight of the steel cables. The Clifden Bridge of 112 m span over the Waiau River in Southland was opened in April 1899 by Sir Joseph Ward. The 7.5-m-high towers are constructed of concrete but are lined out in plaster to suggest stonework. This bridge has a pleasing setting and after the First World War had a Roll of Honour, attached to one of the west towers, commemorating the fallen and also the returned servicemen of the Clifden district. When a reinforced concrete bridge was built in 1978 the suspension bridge was given to the New Zealand Historic Places Trust by the Wallace County Council. Now a footbridge, it was designed by the County Engineer, C.H. Howorth.

In 1899 a Wellington hardware merchant named John Duthie, popularly known as the 'Iron Duke', had a large timber house called Balgownie built at Naenae. Requiring a good water supply, he built a pumphouse nearby in concrete where he installed a steam-operated pump to bring water a considerable distance from a storage dam on the hills above the house. It was said that he introduced a supply of water into the lining of a cool room to maintain a uniform low temperature.

The pumphouse is rather ornate with battlements and thick concrete walls. As the building lacks reinforcement the subsequent ground settlement, and possibly earthquakes too, have caused major cracking with one end wall showing signs of collapse. For many years it has been covered in ivy.

Already mentioned is Ewer's hop kiln at Upper Moutere. Dating from *c.* 1900, it has the ubiquitous corrugated galvanised iron roof. The timber wing abutting against the concrete drying tower was a common feature.

CLIFDEN BRIDGE, WAIAU RIVER.
This 112-m-span bridge has 7.5-m-high concrete towers. They are lined out in plaster to suggest the local limestone. A Roll of Honour was attached to the west tower as a First World War district memorial. The bridge was opened in April 1899.

RACES, WOOLSHED, WHITEROCK STATION, NORTH CANTERBURY.
The woolshed was built in 1891, so these races date from this time. They are rare, if not unique. The shed is supported on tapering concrete piles, foreshadowing their common usage from the 1930s.

Below left: FOUNDATIONS, CROWN BATTERY, KARANGAHAKE.
In July 1889 the world's first field test of the MacArthur-Forrest potassium cyanide process of gold extraction was carried out on this site. A new battery was erected 1892–93 and extended to 60 stamp heads by 1898.

Below right: FOUNDATION RUINS, VICTORIA BATTERY, WAIKINO.
Built by the Waihi Gold Mining Company in 1897, the Victoria Battery began with 100 stamp heads. Another 100 stamps were added a few years later to make this one of the largest batteries in the world. Today the site is a reserve, with Department of Conservation interpretation of the relics.

KOPUA VIADUCT, NEAR ORMONDVILLE. Better known as an architect, G.A. Troup designed several railway viaducts on the Wellington–Napier line including this structure. It was completed in 1897 and has concrete abutments and end span piers.

Chapter Seven
The twentieth century: The first decade

From this time on concrete came into increasingly widespread use and examples cited show the general development. There was of course a significant increase in demand with the introduction of reinforced concrete or, as it was commonly called, ferro-concrete. We have already traced the overseas discovery and early development of reinforced concrete. It was not delayed in finding its way to New Zealand. There had been several pioneering attempts in the latter part of the nineteenth century to introduce iron into the matrix. Presumably the earthquake-resistant properties of a reinforced concrete structure was a potent reason to encourage its use.

One of the most dramatically sited bridges in the country spans the Shotover River at Skippers in Central Otago. A suspension bridge of 96.3 m span was opened on 1 March 1901. It is almost the same distance (91.4 m) above the turbulent river. The banks are precipitous schist rock with the concrete towers rising 11.5 m from them — a rather unusual material in Central Otago, where the towers of so many of the earlier suspension bridges were built of local schist. Fourteen wire cables pass over the top of the towers instead of through them as was sometimes done. Transporting the materials for the bridge over the notoriously fearful Skippers Road must have been a difficult and hazardous task to say the least. As was the custom, the opening was marked by a grand banquet and ball with the guest of honour being the Hon. James McGowan, Minister of Mines.

An early concrete flour mill was built in the Southland town of Otautau in 1902 by William Saunders to replace the 1884 mill destroyed by fire. Power was supplied to the new four-storeyed Otautau Roller Flour Mills from the Wairio Stream nearby. Although milling had ceased and the mill had been without the end wall for some years, it was worthy of preservation. In 1987, however, it was wantonly destroyed as filling for a stopbank. It should have been left as a static industrial monument — a relic of the once widespread flourmilling industry.

Proclaimed at the time as the first reinforced concrete arch bridge in the colony, Dunedin's George Street Bridge is certainly a landmark in the history of bridge design in New Zealand. It is believed that the designer was H. Langevard of the city engineer's staff — the August 1902 contract drawings show his signature. The bridge was completed early in 1903. The structure consists of a reinforced concrete span of 12.19 m springing from solid concrete abutments. These are faced in ashlar masonry as are the wing walls. Ornamental pedestals capped by wrought iron scrolls and handrailing of wrought iron keep this bridge in the tradition of the masonry arch, and the expectation that public structures should make concessions to time-honoured materials. This attitude was common all over the world with the earliest reinforced concrete arch bridges. The George Street Bridge is recognised by the New Zealand Historic Places Trust as a pioneer example of engineering technology.

Manager's house, John Wilson and Company Ltd, Warkworth.
Sited close to the cement works a few kilometres from Warkworth, this house was built in 1904. After the closure of the works in 1928 it became vacant and partly ruined by vandals. More recently it has been made into an attractive and habitable residence.

George Street Bridge, Dunedin.
This pioneer structure was completed in early 1903, being the first reinforced concrete arch bridge in New Zealand. It was designed in the city engineer's office in the previous year. As with its contemporary arch bridges in urban settings elsewhere, it retains heavy ornamental pedestals and stone facings on the abutments in lieu of a more honest expression of concrete.

FARM DAIRY, MARAEKAKAHO STATION, HAWKE'S BAY.
This is an early (1905) use of concrete for such a building. At first the roof was shingled — a somewhat strange combination with concrete. The interior was designed to be hosed down and has concrete benches.

17 PATRICK STREET, PETONE.
Under the Workers' Dwellings Act 1905 the state built houses in the four main centres. In 1908 at the Heretaunga Settlement two houses were built in concrete for a comparison in cost with similar-sized timber dwellings.

Although a mass concrete core was used in 1881 on the 19-m-high Abbeystead Dam near Lancaster, the general British practice until early this century was to use a minimum of concrete or none at all in the masonry core and to face these dams in masonry. The first British mass concrete dam using timber shuttering instead of stone cladding was the Blackwater Dam in Scotland, completed in 1909.

Dating from the period 1904–8 a mass concrete dam was built across the Brook Stream in Nelson. This city council venture was carried out entirely by manual means — the concrete was hand mixed, hand lifted and hand placed — certainly the usual method for early concrete work before mechanical equipment took much of the drudgery and human fallibility out of the process. There must have been a fault in the proportions of the mix and it was probably deficient in sand quality because the too lean a mix subsequently gave considerable trouble with leaking. About 1962–64 an attempt was made to improve matters by reducing the height of the dam. Obviously this wasn't successful and a further reduction was made in the mid-seventies. This time the middle portion was lowered to form a spillway. The original height was approximately 13 m and the length of the straight dam is 88 m.

It was most appropriate for the manager of John Wilson and Company Ltd to live in a concrete house provided by Nathaniel Wilson and close to the cement works near Warkworth. In 1904 Nathaniel built a substantial house in reinforced concrete for his son as manager of the works and it was used as such until the works closed in 1928. For many years thereafter it was vacant and became a ruin after vandals, arsonists and then the elements took their toll. More recently it has been fully repaired and made habitable — a remarkable transformation which reveals its impressively solid concept. Single storey, it has a generous basement and a twin-gabled verandah with a concrete slab floor and precast concrete balustrade. The roof was of Marseilles tiles. Concrete surfaces are plastered on both faces. This house had the distinction of being the first in the district to be lit by electricity.

Designed in 1899 the Cathedral of the Blessed Sacrament in Christchurch was completed in 1905, being very highly praised as one of the finest churches in the country. It was the work of F.W. Petre, who continued his fondness for neoclassicism in this High Renaissance example. The scale and massing of the elements is certainly impressive. In keeping with tradition the building is faced in stone, both inside and out, on a concrete structural core.

An early use of concrete for a farm dairy is that at Maraekakaho station in Hawke's Bay. It was built in 1905 and was part of a large complex of buildings of which several have disappeared over the years. Floor and walls are of concrete with a timber-framed roof, originally shingled but now sheathed in corrugated iron. The combination of concrete and wood shingles was an odd one. Along the rear and end walls of the interior runs a continuous concrete bench in two tiers with round concrete legs for ease of hosing down. The slab floor extends along the front of the dairy with a drainage channel leading to a sump so the utensils could be cleaned efficiently. The building was well ventilated, having louvred windows on all four sides to ensure cross-currents of air.

The first state housing scheme initiated under the Workers' Dwellings Act of 1905 provided houses predominantly of timber construction in the four main centres. However, at the Heretaunga Settlement in Petone two houses were designed in ferro-concrete, along with similar-sized houses in timber. Built in 1908 they were single-storey with the exterior concrete walls 225 mm thick. With the aggregate close at hand, thus keeping costs reasonable, it was found that the concrete houses cost only £6 more than the timber ones, and they were heralded as requiring less maintenance. As with the other 26 houses in Patrick Street they are still in use and the group has recently been declared a conservation area.

In 1905 an interesting and important Wellington-based publication appeared with the name *Progress — Incorporating The Scientific New Zealander*. As a monthly journal it not only reviewed the latest advances in technology overseas but kept its readers informed on building and engineering developments throughout New Zealand. Some of the articles on concrete design and construction were stimulating and aroused considerable interest with occasional controversy. There were also cautionary tales such as the one in the issue of May 1906 headed 'Shear and Adhesion in Reinforced Concrete' which stated there was a need for care by workmen and competence by architects and engineers; although enormous strides had been made on that comparatively new material in the last two years, there was a need for good adhesion of plain bars. The article recommended further investigation into the use of diagonal bars to take the shearing stresses in reinforced concrete.

In May 1908 an article by C.A. Lewis (described as an eminent architect) commented, with what seems today to be considerable bias, on the failure of ferro-concrete. In the following month's journal there was a rebuttal of this by an engineer stating that if properly designed and executed there were no problems. The September issue produced another rebuttal by C. Fleming Macdonald, an engineer and building contractor of Dunedin.

The June 1910 issue of *Progress* stated that the Society of Scandinavian Portland Cement Manufacturers had reported to the International Association for Testing Materials on sea water tests of Portland cement. A committee, appointed in 1896 to investigate this subject, had found that there was a gradual and regularly increased strength even up to the ten years' test and concluded that the action of sea water alone was not able to destroy the mortar, as the Portland cement appeared to be more or less unaltered.

Bridges have been of the greatest importance in the development of New Zealand, and especially so in rural areas, from the early period of settlement. Fords and punts were no substitute for bridges, being hazardous in times of floods and involving lengthy delays. By the end of the century a large volume of replacement bridging was added to the lists of unbridged sites still requiring attention. Reinforced concrete showed early promise of providing more acceptable standards of design and construction. Hitherto timber truss and trestle bridges decayed at a surprisingly rapid rate, particularly in the wetter districts, whereas steel in the coastal areas corroded all too quickly by rusting.

The development of reinforced concrete structures was given a remarkable boost for local bodies when the Taranaki County Council decided on a policy of replacing existing bridges with the newer material. The decision was based on the experience gained from some early trials that proved these bridges would be cheaper in capital cost, quicker in erection time, and promised a longer life with far less maintenance work and cost. It was found that concrete was an ideal material to supersede both timber, which rotted quickly in the rather damp Taranaki climate, and steel, which rusted rapidly in the heavily salt-laden air. The completion of the 57.9-m bridge over the Waiwakaiho River on the northern side of New Plymouth in late 1907 was heralded as producing the longest reinforced concrete bridge in the country. At that time the county had built nine bridges in this material. They varied in type of design — some were simple beam and slab on piers and abutments, others were concrete arches, and one used the knee brace or understrutted method.

Throughout the history of civil engineering in communication and transport structures in New Zealand, the relatively

BRIDGE OVER WAIONGONATI STREAM, TARANAKI.
Located on Davis Road Extension this bridge was built between 1908 and 1912, and is one of several structures designed by E.C. Robinson.

BRIDGE OVER NGATOROITI STREAM, TARANAKI.
This is on Bedford Road and similar to that over the Waiongonati Stream. It also dates from 1908 to 1912 and was a Robinson design.

early and successful use of concrete revolutionised bridge building. The bridging of rivers was always one of the most demanding and frequent tasks of both central government and local bodies. It is therefore quite extraordinary that the Taranaki County Council was able to embark on this lengthy project using the design and supervisory skills of its engineer, E.C. Robinson, rather than a well-established consultant engineer.

An interesting small bridge was constructed in 1907 in Tariki Road over the Manganui River using spans of 6.4 m, 19.3 m and 8.2 m with piers 9.75 m above the riverbed. It has struts on a rake for the central span to give the knee-braced effect. The thrust against the piers is restrained by horizontal ties. The only alteration to this bridge has been new handrails installed in 1950–51. It was passed over to the Inglewood County Council in 1920.

In Radnor Road there are several small concrete arch bridges dating from the first decade of this century. They were transferred to Stratford County following boundary changes many years ago, but there are no records available of their precise dimensions.

Although the 1907 bridge over the Waiwakaiho River, mentioned earlier, has been demolished, it is of interest to note that it had two spans of 18.2 m and two others of 9.1 m with concrete piers and abutments. A contemporary account in *Progress* for February 1908 stated that the design embodied a combination of Winch and Hennebique construction using flat-topped arches broken into ribs by light wooden frames that remain embedded in the concrete. The reinforcement was parallel rows of 40-pound (18.14 kg) steel rails running horizontally along the whole length of the bridge about 6 inches (150 mm) from the deck level and the same number of 40-pound rails following the curve of the arches at about 6 inches from their lower surface. The Portland cement concrete mix had principally beach gravel in differing proportions. The ring of the arch was 4:1 and all other parts 6:1. The concrete was made very wet and was well spaded and rammed around the steel. The cost was very much less than for a steel bridge, the paper reported.

The Taranaki County Council continued to build concrete arch bridges in the period 1908–12. The Waiongonati Stream is crossed on Old Mountain Road (now Davis Road Extension) by a 13.5-m earth-filled reinforced concrete arch. This bridge, of 16.7 m in length and now disused, is interesting in having a marble plaque with metal lettering (the latter regrettably removed by vandals) set into the solid balustrade to form a milestone — a very rare marker in New Zealand. The bridge dates from 1910–12.

A similar concrete arch spans the Waipuku Stream on Old Mountain Road near the present State Highway 3. Built in 1907 it has a span of 14 m and is 16.2 m in length with low, solid balustrades. An inset stone records that the engineer was

WHEY TANK STAND, PIAKAU CREAMERY, TARANAKI.
This survivor from the creamery was built c. 1910 on the corner of Bedford and Durham Roads. For early reinforced concrete, it is surprisingly refined in its proportions.

E.C. Robinson and F.M. Grayling the contractor.

There are four small early bridges in Bedford Road built by the Taranaki County Council. The largest of the arch bridges is of 15.1 m span across the Ngatoroiti Stream. The dates of these bridges are not known. With the exception of the Radnor Road bridges and the Waiwakaiho Bridge, those mentioned were transferred to the Inglewood County in 1920 when a change of riding took place.

Early reinforced concrete in Taranaki was not confined to bridges. About 1910 the Piakau Creamery, a branch of the Maketawa Dairy Factory, was built near the junction of Bedford and Durham Roads, 8 km from Inglewood. It had concrete foundations and floor with a loading platform and a concrete tank for cooling water. The lower portion was used for storing butter as a service to local farmers. Alongside there is a free-standing open stand which supported the whey tank. It is of finely proportioned reinforced concrete with posts carrying haunched beams to resist shear stress. The two internal posts are corbelled on all faces. Diagonal braces on the four sides complete this elegant structure.

A small stream alongside was dammed with a concrete wall to retain a head of water for flushing waste products downstream — a practice that would not be tolerated today. The creamery closed about 1936 and the superstructure was removed.

The year 1907 was a significant one for the establishment of reinforced concrete in Auckland. The November issue of *Progress* reported that ferro-concrete and armoured concrete were one and the same thing and that its first use in Auckland was in the harbour board wharves. The Ferro-Concrete Company of Australasia under its Engineer-in-Chief, Mr Moor (*sic*), and an engineer, W.H. Hamer, was the first in New Zealand to undertake construction and erection of reinforced concrete structures using the Hennebique system, it stated, adding that great care was being exercised in the details. *(To page 112)*

No. 5 Pumphouse, Waihi Gold Mining Company, Waihi.
A striking industrial monument, the pumphouse was built in 1904 to house the giant steam-driven pump used for raising mine waters to the surface. Later the pump was electrically operated. Abandoned since the Martha Mine closure in 1952, the pumphouse stands as a gaunt reminder of mining technology at the turn of the century.

CEMENT WORKS RUINS, NEAR WARKWORTH.
The ruins of various concrete structures erected from the 1880s to the early 1900s have produced some dramatic visual images.

TARIKI ROAD BRIDGE, MANGANUI RIVER, TARANAKI.
The Taranaki County Council was very quick to see the advantages of reinforced concrete for bridges. Built in 1907 to a design by E.C. Robinson, County Engineer, this example has understrutting in concrete, hitherto seen in some early timber bridges.

Opposite: SKIPPERS BRIDGE, UPPER SHOTOVER RIVER, CENTRAL OTAGO. Replacing a lower level bridge of 1868, this is the highest suspension bridge in the country at 91.4 m. Opened in early 1901, it has concrete towers to support the cables carrying the deck over a span of 96.3 m.

GOVERNMENT BATH HOUSE, ROTORUA.
Opened in 1908 and now known as Tudor Towers, this building has been a focal point in the Government Gardens. The Elizabethan style, unusual for a spa, shows a timber frame with precast pumice concrete panels — an early example of such work.

Opposite: CATHEDRAL OF THE BLESSED SACRAMENT, CHRISTCHURCH.
This imposing edifice was designed in 1899 by F.W. Petre and is generally regarded as his most important work. It was completed in 1905. The concrete structural core is clad on both faces in stone, in keeping with the High Renaissance classicism.

CYANIDE TANK BASES, WAIKINO.
In 1907 the Waihi Gold Mining Company constructed interlinked concrete bases to support the array of 32 steel B & M cyanide tanks. Removal of the tanks for scrap metal after closure in 1952 has revealed intriguing sculptural forms.

Works then being carried out by this firm included the Railway Wharf under construction, Queen Street and ferry service wharves, Admiralty jetty, foundations for No. 1 crane at Calliope Dock, and two five-storey buildings, one being a mill for the Northern Roller Milling Coy and the other an adjacent block, for one Lichtenstein, on 15.2-m-long piles driven into rock on reclaimed land. The only woodwork in the mill was in the doors and window frames. Other buildings included a block of offices in Swanson Street and a block of shops in Queen Street, as well as a house in Remuera. Grafton Bridge was also listed.

The mill for the Northern Roller Milling Coy was in fact a storage building alongside the former brick Eight Hours Flour Mill of J.C. Firth, which business they acquired. The 1907 building had a most unusual construction in that it appeared to have been detailed using timber sizes and practice. For example, the king post roof trusses were of ferro-concrete as well as the open string staircase. This structure was demolished in the 1960s to make way for the new Customhouse.

To illustrate the readiness of New Zealanders to accept reinforced concrete construction for buildings, we note that the first such building in New York was the Schirmer factory, built in 1905. In San Francisco the disastrous earthquake of 1906 encouraged the use of reinforced concrete in the replacement of several demolished buildings soon afterwards.

However, the architects were loth to appreciate the true merit of concrete as a legitimate building material other than for structural purposes. They proceeded to cover it with various facings, and this attitude even extended to Frank Lloyd Wright, who was noted for his individualism in architectural philosophy.

In Britain only one non-industrial reinforced concrete building did not have a cladding disguise in the first decade. A factor inhibiting the adoption of reinforced concrete in England at this time was the difficulty architects had in designing concrete structures to comply with the very conservative building regulations.

The first visual evidence of concrete on the Waihi goldfield was the erection in 1904 of the prominent No. 5 Pumphouse for the Waihi Gold Mining Company at its renowned Martha Mine. This was a handsome, chapel-like structure measuring

GRAVING DOCK, PORT CHALMERS.
Begun in 1905, this project, designed by A. Luttrell, was not completed until June 1909. The sculptural forms of the gothicised archways over the several access steps increased the strong visual impact of the structure. Regrettably, although still extant, it cannot be seen as it has been interred under the container terminal.

15 m by 9.1 m on a very substantial foundation. Initially steam driven, it was later converted to electric operation when the company installed its own hydro-electric power station at Horahora on the Waikato River. After closure of the Martha Mine in 1952 the pumphouse became derelict — a silent, forlorn but nevertheless imposing industrial monument. Recently a new company, now mining behind the pumphouse, has cleaned an extensive area round it, making public access into the building possible. As an 'Engineering to 1990' heritage structure honoured with a commemorative plaque, it arouses a lot of interest today.

Also in Waihi, not far away on Union Hill, are the monumental cyanide tanks erected for the Waihi Gold Mining Company about 1905. Known as B & M tanks after their designer, C.F. Brown and his partner McMiken, they are also referred to as tall tanks and in some countries as Pachuca tanks. Built of concrete they are 16.7 m high and form a cluster of six. The lower portion of the cylinder has an inverted cone which sits on a base. Although no longer used these tanks are an impressive reminder of the more spectacular developments in goldmining technology.

C.F. Brown invented the agitation cyanide tank when he was general manager of the Komata Gold Mines and he tried out his first tank, made of steel in 1902, at his own site. This form of cyanide tank has a central pipe which passes a charge of compressed air into the pulp of crushed ore and potassium cyanide solution.

With the setting up of the Department of Tourist and Health Resorts the government appointed the balneologist Dr Arthur Stanley Wohlmann to a new position in 1902 at Rotorua. He had a vision for a grand new bathhouse that would have many facilities for treating patients suffering from rheumatism and arthritic conditions.

Planning began in 1903 under B.S. Corlett, Inspector of Works for the department and W.J. Trigg, Public Works Department draughtsman, with a local architect, J.W. Wrigley. The Elizabethan style chosen made use of a rimu frame with the main beams of totara, and kauri for the joinery. The infill is of pumice concrete (precast) and solid concrete was used for 1.8-m-high arches in the basement that give access for the many pipes and drains required in a spa building.

A very considerable quantity of pumice concrete was used in the construction. The bathhouse was intended to be a major attraction and much attention was given to the design of the public rooms and an impressive foyer. However, the highly acidic mineral waters took their toll on the plumbing and, with the steamy conditions, interior wood-pulp plaster and paintwork suffered severely. Indeed the very high maintenance costs caused the department to lose its initial enthusiasm for the bathhouse. It was opened in 1908 and ceased to function as such in 1963 when the Rotorua City Council took over responsibility for the building and converted it to a museum. For the next three years some bath treatments continued to be administered. The building was given the name Tudor Towers and still makes a handsome focal point for the extensive and attractive Government Gardens.

A considerable stimulus was given to the construction industry in the South Island with the arrival (c. 1901–2) of the Luttrell brothers in Christchurch. Born in Australia, both Alfred and Sidney had spent most of their lives in Tasmania where Alfred had set up practice in 1886. He was an architect and engineer, and in 1897 Sidney joined him in partnership as the business and contract manager. There is little evidence of any concrete construction by the firm in Tasmania apart from foundations. In New Zealand the Luttrell brothers specialised in reinforced concrete commercial buildings and grandstands for racing and trotting clubs. They also had a number of commissions from the Roman Catholic Church.

The most probable first use of reinforced concrete for commercial work in Christchurch was their building for the New Zealand Express Company. A tall structure for the time, it rose 39.6 m, reinforced concrete being limited to the foundations and the two lower floors. Above them structural steel was used. This brick-faced building was almost complete by June 1906 but it was not in use until the following year.

Port Chalmers saw a start made in 1905 on a large concrete graving dock. With a length of 162.7 m, it has the usual stepped sides and ends known as altars. Strong sculptural qualities were evident with the access ports having quaint Gothic arches over the flights of steps. The designer was Alfred Luttrell. Construction was attended by various problems and frustrations with the Otago Harbour Board as the sixteen months estimated for the contract dragged on to three and a half years. Before the completion in June 1909 the Luttrell brothers had handed over to Scott Bros. Like its masonry predecessor of 1872, this dock still exists but cannot be seen, for the demands of container marshalling space caused the board to fill in and inter two splendid industrial monuments.

NEW ZEALAND EXPRESS COMPANY BUILDING, DUNEDIN. Now known as the MFL Building, it was designed by A. Luttrell and completed in 1910. A pioneer structure in building technology in New Zealand in the use of precast concrete slabs, it sat on a raft foundation to counteract poor ground conditions. It is a splendid structure architecturally.

Alfred Luttrell designed a second building for the New Zealand Express Company. This was in Dunedin, beginning in 1908 with a contract awarded to C. Fleming Macdonald. Contract completion was achieved in 1910. The tallest building in the city, it was a pioneer in concrete technology. It was the first time in New Zealand that precast concrete slabs made off the site had been used. (These slabs should not be confused with the precast concrete blocks used on the 1882 lighthouse on Burgess Island in the Mokohinau group.) The seven-storey structure has reinforced concrete raft foundations to overcome the problems of building on the old foreshore. The frame is of reinforced concrete. Architecturally it is quite a splendid building, sensitive in scale and detailing of the elevational elements and adding vitality to the streetscape. It justifiably has a high preservation rating from the New Zealand Historic Places Trust. It is also a rarity for a commercial building in this decade, being unadorned by exterior cladding of brick or stone.

The first reinforced concrete building in Wellington dates from 1907 when construction began in Panama Street for Sefton Moorhouse. Designed by Charles Tilleard Natusch, it was of four storeys and had Kahn bars used principally for the beams, with expanded steel for walls, floors and the flat roof. In the same year concrete retaining walls were being erected in Kelburn, the architect being John S. Swan.

In the following year an announcement was made of Atkins and Bacon's design for the Nelson Moate and Coy building on the corner of Wakefield and Blair Streets. This is a three-storey reinforced concrete building with a basement. It was designed for heavy live loads, as well as earthquake and fire resistance, in column and beam construction. It is still in use.

Also in the capital in 1908 a start was made on building the upper dam at Karori. This is a 21.3-m-high gravity type, but of arch form, designed by W.H. Morton, City Engineer. He was also responsible for the reinforced concrete chimney of the city destructor, no longer in existence. It was claimed as the first reinforced concrete stack in the Southern Hemisphere and although *Progress*, April 1908, does not give the height it states that the foundations consisted of 42 piles supporting a 6-foot

Upper dam, Karori Reservoir, Wellington.
Completed in 1910, the dam was designed by W.H. Morton, City Engineer. It is a gravity dam, although arched. In recent years it has been dewatered and is to be part of the Karori Wildlife Sanctuary.

(1.8-m) bed of reinforced concrete. The reinforcement of the chimney had vertical bars approximately 30 feet (9 m) in length with substantial overlap and bound together by rings of round steel bars at 18-inch (450-mm) intervals. The thickness of concrete varied from 12 inches (300 mm) at the bottom to 6 inches (150 mm) at the top.

Another tall chimney had been built in 1903 at Wilsons' Portland cement works near Warkworth. This was 34.74 m high of reinforced concrete and lined with firebricks for 15.24 m. The cornice was coke breeze (a light-weight concrete). This chimney was reputed to be the first to be built of this material.

For office buildings of the first decade there was still some reluctance among architects to dispense with brick or stone facing to concrete, and no doubt some clients would have asked for the traditional materials. In 1909 a five-storey building was begun in Wellington on the corner of Willis and Mercer Streets. Known as Cooper's Building, it was designed by J.S. Swan in reinforced concrete frame with a flat roof, and pressed bricks on the street elevations. It was used until the late 1980s when it was demolished after a fire.

Early 1909 saw a start on pile driving for a four-storey office building in Brandon Street, Wellington, for Chapman, Skerrett, Wylie and Tripp, a legal firm. The architect was Frederick de Jersey Clere, whose earliest essay in concrete was the

CHAPMAN, SKERRETT, WYLIE, TRIPP BUILDING, WELLINGTON.

Begun in 1909 this was designed by F. de J. Clere for a legal firm. The brick-faced elevation and the large segmental arches at ground-floor level make an attractive facade. However, the later verandah awning cuts across these in an unfortunate manner.

1884 country house Overton, near Marton. His specification for the piles of the office building was one part of Crown brand cement, two parts of sand, four parts of washed gravel. The piles were stacked for seven days after being made and were driven to a depth of 3.6 m. They tapered from 300 mm by 300 mm at the foot to 250 mm by 250 mm at the top. The calculated strength for compression was 44.2 tonnes but the force actually exerted by the impact of the ram on the head of the piles was 60.45 tonnes — 16.25 tonnes in excess of the calculations. The Clerk of Works was G.J. Bertinshaw, a specialist in concrete work. The Crown brand of cement specified was made by the New Zealand Portland Cement Company, which amalgamated with Wilson's Portland Cement Company Ltd in 1918 at Portland. The very attractive front elevation of this building is brickfaced, with rusticated columns and large ground-floor windows with segmental arches, and main entrance. A later verandah awning cuts across these features.

It was about the beginning of 1909 that the proprietary name 'Camerated Concrete' appeared in New Zealand with advertisements in *Progress*. This form of concrete was claimed to be suitable for any kind of building, 20 percent cheaper than brick, 5 percent dearer than wood and with full particulars being available from W. Leslie Friend, Auckland and Christchurch. By 1912 the advertisement stated that it was the cheapest permanent construction known in the building world, being fireproof, damp proof and resistant to earthquakes. An extract of a letter to *Progress* received 15 October 1911 after an earthquake in Napier stated:

> There was not the slightest damage done to any of the buildings erected or partly erected in Camerated Concrete by the recent earthquake . . . The earthquake was one of the severest we have had for many years (damaging numbers of brick buildings). Architects are now quite satisfied with the stability of Camerated Concrete and we are confident that the success of the system is assured.

This method received the support of such well-known architects as Wilson & Moodie, Auckland; John T. Mair, Wellington; F.de J. Clere, Wellington; and F.C. Daniell, Hamilton, so it must have been considered thoroughly satisfactory at the time. It was a system patented in 1905 by Henry A. Goddard in Australia and consisted of walls with inner cavities formed with removable steel cores.

Grafton Bridge was mentioned as one of the works being produced in 1907 by the Ferro-Concrete Company of Australasia. In February 1910 *Progress* described the completed bridge thus:

> This bridge is the largest concrete bridge in the world. One central span with two approaches of which the western is of two spans of 35 feet [10.68 m] and four of 75 feet [22.8 m] and the eastern is of three spans, two being of 83 feet [25.29 m] and one of 42 feet [12.8 m]. The bearing plates for the girders are fixed on the piers, with ample room for expansion. The main piers completing the approaches stand 100 feet [30.48 m] and are cylindrical, the walls being from 12 inches [300 mm] to 8 inches [200 mm] thickness. On the top or head of the piers curved cantilever brackets carry the footpaths over the piers, and serve to embellish the pier heads. The piers are really built on three walls, and between the walls sit the abutments, being entirely independent of the piers. Into the abutment is fixed a thrust plate or steel grillage set to an angle normal through the line of thrust, and upon this plate rests a hinge supporting the whole arch.
>
> The arch span is 320 feet [97.6 m] and is three hinged. It consists of two ribs ranging from six feet [1.8 m] at the abutment to nine feet six inches [2.9 m] graduating to five feet [1.5 m] at the crown. These huge ribs are tied together by beams which act also as wind braces.

CYANIDE TANKS, UNION HILL, WAIHI.
Known as B & M, tall, agitation, or Pachuca tanks, these tanks were a New Zealand-patented development. The design was for steel tanks, and this cluster of six is the only example in concrete. The tanks were built c. 1905.

The height from the deepest part of the gully is 147 feet [44.8 m], the width of the structure is 37 feet [11.27 m] overall, including the roading and footpaths, the road being 24 feet [7.3 m] and the paths 12 feet [3.6 m] each.

The parapet is concrete 4 ft 9 in [1.44 m] and carries twenty-six lamp posts, lit by electric light. The curbs of the paths are of Coromandel granite, and the grade of the bridge is one in seventy-four, with a total length of 950 feet [289.5 m].

The superstructure consists of T-shaped piers, this form being selected for its lightness; the longest pier is 80 feet [24.3 m] and the whole is braced by two T-shaped beams. These piers are constructed on arched rings at 21 feet [6.4 m] centres, and support main girders, the girders carrying the secondary beams of the roading and footpaths.

The decking of the roadway is six inches [150 mm] thick, and will receive a layer of inch and a half [38 mm] Neuchatel asphalt. The footpaths are four inches [100 mm] thick and will receive a finishing layer of inch and a half [38 mm] concrete.

223 Ettrick Street, Invercargill. Built *c.* 1910, this is one of many houses erected in Invercargill using concrete blocks. The initiative for producing such blocks seems to have come from the Invercargill prison authorities.

    The bridge . . . is to be opened early in April. The whole construction is of reinforced concrete throughout, and it is interesting to note that in the approaches it is the longest continuous reinforced girder extant, while the span is the longest so far constructed of reinforced concrete. The Dominion, therefore, once more heads the world.
    The work was done under the superintendency of the Auckland City Engineer, Mr W.E. Bush and the cost (estimated) was £39,480.

Designed by R.F. Moore, Chief Engineer of the Ferro-Concrete Company of Australasia, Grafton Bridge has the highest classification given by the New Zealand Historic Places Trust, in recognition of its outstanding technological merit and its magnificence as a townscape element.
    Occasionally strange but visually exciting forms are produced in concrete and, where partial demolition has taken place, they can be the result of fortuity. At Waikino the Waihi Gold Mining Company installed in 1910 a close-knit group of 32 B & M or agitation tanks used for potassium cyanide treatment of gold-bearing ores. Built of steel cylinders some 15 m tall,

347 Ettrick Street, Invercargill.
One of the many concrete block houses built from 1909 to the early 1920s in this city. A local company produced a catalogue of plans and photographs, with estimates and specifications to encourage custom.

91–93 Venus Street, Invercargill.
These are two of a group of four similar houses, built *c.* 1910 with a concrete surface devoid of plaster. Such 'brutal' treatment seems out of character for the period.

they were supported by circular concrete structures having six round arched openings and on hexagonal plinths. The arched openings allowed pipes to be run with access to the conical bases of the steel tanks. The tops of the tanks were removed for scrap metal after the closure of the Martha Mine in 1952, and the bases now provide intriguing and dramatic abstract sculptural elements enhanced by their honeycomb pattern. They also constitute a very important visual relic of mining technology.

A rather commonplace object for inclusion in this survey of concrete usage is the fence post. Although a familiar enough sight now, an interesting early example is that used by the New Zealand Railways to fence the Midland Line in Canterbury, where it passes through high country runs in the Avoca–Cass section. In addition to the railway line there is a fence on the other side of the road leading to Craigieburn sheep station. The dates for the fence erection approximate that of the progress of the railway. It reached Broken River in 1906 and Cass in 1910 so that the Craigieburn section would be about 1909. The posts are unusual in section, being T-shaped and having a cap — some signs of rust indicate that there was reinforcing.

It is relevant to note that a 1911 report of a study made by the American Railway Engineering and Maintenance of Way Association on concrete fence posts states:

> Indications are that they are practicable, durable and economical. Those with a cross section approaching a square form are easiest made and most suitable for fencing. Spacing should be the same as for wood posts. Sixty days curing in the yard is recommended for best results.

At the end of the first decade the Post and Telegraph Department in New Zealand was testing ferro-concrete telegraph poles. An October 1910 report in *Progress* described the dimensions as being 100 square inches (645.16 sq. cm) at the base and 37 square inches (238.71 sq. cm) at the top with a total length of 36 feet (10.9 m) and 2,600 pounds per square inch crushing load (17.926 kPa). There was also reference to testing carried out by the London Post Office. It was claimed that in nine years concrete poles would be seven times stronger than ironbark and fourteen times stronger than totara.

Progress had also reported in its October 1906 issue that 'rapid progress has been made over the past five years with a new material — the hollow concrete block . . . the industry has grown almost as surprisingly as the manufacture of Portland cement.' These blocks were practically unknown in 1900, *Progress* continued:

but it is probably safe to say that at present more than 2,000 companies and individuals are engaged in their manufacture in the U.S. and the United Kingdom. The cause in the U.S. is the rapidly failing lumber supply and the widespread interest in all applications of Portland cement.

There is no mention of any New Zealand use of such blocks. The first patent for concrete blocks (presumably solid) was issued in 1875 in England. It seems that concrete blocks imitating masonry were in use in the United States from 1868 when one Geary patented his design. We have already noted the use of purpose-made blocks for Mokohinau light in 1882. In his *Building Materials of Otago*, published in 1870, W.N. Blair states that in England concrete blocks were made in large quantities by machinery for arch-stones, quoins, sills, lintels, steps and mouldings of all kinds — but not for general building.

We have mentioned the construction in concrete of New Plymouth Prison in 1877. It is of interest therefore to observe that in 1908–10 a new Invercargill Prison was built of concrete with all construction being carried out by the inmates. In his 1909 annual report the jailer stated: 'The quality and quantity of the work performed by the prisoners on this prison proves conclusively that under proper management prison labour can be utilised in almost any direction.' In his report for the following year he mentions that a concrete-blockmaking plant had been installed and orders were coming in: 'This will prove a very profitable industry and provide a means of employment for boys. I hope that before long a cement-pipe making plant will be added to this industry.' A photograph in the Public Works Statement shows the large prison yard stacked with concrete blocks awaiting delivery.

The importance of this local industry can be seen today in the number of concrete block houses built in Invercargill from that time and into the 1920s. The prison certainly had a fine sample to show intending users in the extensive boundary wall to the south of the institution. It presents a rusticated surface on the exterior — a treatment to give an impression of stone sometimes referred to as 'cyclopean' or 'rock-faced'.

About this time the Property & Finance Co. was established in Invercargill to promote home building from a catalogue of plans and photographs of completed houses. Estimates of costs and specifications were available. Although some designs were described as being suitable for brick, concrete or concrete block construction, concrete block was the most popular. A good example still stands at 109 Earn Street, built in 1909 probably before the prison blocks were available. It is a two-storey house built by a man named Kennedy who made the rusticated finish blocks on the site. The fence is also of these blocks as were many others.

One house built a little later is that known as the Thompson House, at 322 Crinan Street. It is a large two-storey residence used for a period as a hostel by Southland Girls' High School but unoccupied for some years. Built in 1922 it uses 250 mm by 250 mm concrete piles within a continuous concrete foundation of filled blocks of 600 mm by 300 mm by 200 mm. The last dimension is the width of the block, which had 50-mm walls and keyways at the ends. These same blocks were used above the foundation and had the rock-face exterior. Presumably these were prison-made blocks.

One of the old type of two-storey corner shops to be found in residential areas can be seen at the corner of Elles and Crinan Streets. It has the same concrete block exterior and a bull-nosed verandah. At 347 Ettrick Street there is an attractive single-storey house in immaculate condition built of rusticated concrete blocks with an octagonal turret-roofed projection. It has an Historic Places Trust classification.

In the same area of Invercargill, in Venus Street, there is an unusual row of four small houses all built of concrete and dating from about 1910. Narrow and rectangular they present uniform street facades, each hipped gable having a timber battened end. The houses are unusual in that they are built of concrete with an 'off-the-form' finish; that is, after the formwork was stripped there was no attempt to plaster the concrete or even to touch up or 'bag' the surface. The line of each concrete pour can be readily seen. Such frank acceptance of raw concrete as a finish in domestic architecture was certainly rare for the time, anywhere in the world.

About the turn of the century John Wilson and Company Ltd carried on a strong public relations drive to market their cement by general publicity and also by introducing plans for houses using their product. In 1910–12 they published catalogues and in Auckland, where the marketing thrust was concentrated, intending home owners responded. A traveller for the firm, Matthew Blair, decided to build his own home and *c.* 1910 he took six months' leave from his employers in order to work on it himself. The house, at 7 Dexter Avenue, is still standing, being a bungalow of five bedrooms with an attic bedroom he added later. There are twin gables flanking a small entrance porch and the roof is of corrugated iron. The concrete surfaces are plastered on both faces and the structural condition today is sound.

Another concrete house was built in the same area about this time, *c.* 1911, at 4 Kingsview Road. It has a double skin with some reinforcing in the space and is plastered both externally and on the interior. There is some fine plasterwork to be seen here, while for the decorative work outside an expanded metal was used to give a good key. At some later stage a small attic was built at the rear. Set well back from the street, this fine house was erected for a Mr Avery. The use of the double skin overcame the problem of condensation so prevalent in Auckland.

GRAFTON BRIDGE, AUCKLAND.
When opened in April 1910 this splendid structure had the largest concrete arch span (97.6 m) in the world. It was designed by R.F. Moore for the Auckland City Council.

Grafton Bridge, Auckland. This detail shows the massive piers at each end of the arch span. The top of the pier has branching cantilevered brackets to carry the footpath. Grafton Bridge was a remarkable achievement for the time.

Opposite: Fence post, Midland Railway, Craigieburn. When the railway was constructed through the high country runs it was necessary to erect fences to keep stock clear. The New Zealand Railways designed their own fence posts in concrete. This section of the line dates from *c.* 1909.

Chapter Eight
1910–19

The second decade of the twentieth century was a period when building and construction in general slowed down very considerably because of the impact of the First World War. Nevertheless there was some significant development in reinforced concrete structures in New Zealand. There was an increasing acceptance of concrete as the most suitable material for commercial, industrial and even public buildings. It was certainly becoming more popular with engineers for bridge design.

F.de J. Clere, whose first effort in concrete goes back to 1884, was responsible for the very attractive Anglican Church of St Mary the Virgin at Karori in Wellington. Dating from 1911 this is an essay showing the Spanish Mission influence in architecture and it is particularly well handled in plastered concrete with a timber ceiling treatment. Successive additions and alterations in 1917 and 1927 were also by Clere and the later additions by Porter and Martin likewise have respected the integrity of the original. From the concrete purist's point of view St Mary's does not seem to advance the understanding of reinforced concrete, but it was nevertheless a comparatively early use of this method of construction in ecclesiastical work.

In 1911 the Taranaki County Council's policy of replacing timber and steel bridges with reinforced concrete structures continued with the Bristol Road Bridge over the Manganui River. It consisted of five spans with four of 15.2 m and one of 9 m, giving a total length of 71.3 m. This bridge has the characteristic solid balustrade of the period.

In addition to the construction of the upper dam at Karori Reservoir, the Wellington City Council decided to build another reservoir at Wainuiomata near the Orongorongo Range. The City Engineer, W.H. Morton, was again the designer and the dam came into use in 1911. Unlike the Karori arch dam, this is a 164-m-long buttress type with a battered face upstream and a series of transverse vertical walls forming cells to support the deck and act as buttresses. It is 12.5 m high. The risk of a major earthquake causing the 80-year-old structure to fail and release a surge of water to the residential areas downstream determined the dewatering of the reservoir in the early 1990s, after the Te Marua Lakes scheme became operative. The Morton Dam (so named after its designer) is an interesting structure and like so many of New Zealand's public reservoirs it is set in idyllic surroundings of bush and hills. In spite of redundancy it joins the ranks of abandoned technological monuments which contribute to the country's heritage of industrial archaeology.

In Auckland, 1911 saw the completion of the city's reservoir at Mount Eden. In the following year in that city, New Zealand-born Robert Wladislas de Montalk designed and built the Hotel Cargen in reinforced concrete. At the time the Auckland Master Builders' Association strongly opposed the use of this method as a newfangled idea and its members engaged in a boycott for a period.

CHURCH OF ST MARY THE VIRGIN, KARORI, WELLINGTON.
This Anglican church was designed by F.de J. Clere and shows the influence of the Spanish Mission style. In spite of three lots of additions and alterations this very attractive church retains architectural integrity.

Bristol Road Bridge, Manganui River, Taranaki. Dating from 1911 this five-span concrete girder bridge was designed by E.C. Robinson, Taranaki County Council Engineer. The solid balustrade was common on early concrete bridges.

However, in 1912 the Auckland Technical College (now the Auckland Institute of Technology) had its four-storey reinforced concrete building in Wellesley Street virtually completed. Measuring 50.7 m by 29.26 m it was designed by architect John Mitchell and the contract was supervised by the well-known architects Edward Mahoney and Sons.

Napier had a building under construction in 'Camerated Concrete' the same year. This was Hukarere School on Bluff Hill, a three-storey block with prominent buttresses.

In Wellington J.M. Dawson's design for a three-storey commercial building on the corner of Cuba Street and Swan Lane was completed and it survives today as a pleasant, well-proportioned structure.

Water supply for the city of Auckland has been noteworthy for the splendid pumping station and its Victorian beam engine at Western Springs. This is a brick structure but the later reservoirs in the central parts of the city, such as those in the Karangahape Road and Khyber Pass areas, are concrete. The Khyber No. 2 Reservoir erected in 1912 is a utilitarian structure, as is to be expected. Sited in a minor street it makes slight concession to the streetscape by way of ornament. The walls are plain with off-the-form surfaces now showing considerable spalling and erosion. Below a parapet is a frieze with a central curved pediment that has a panel with the name and date in plaster relief lettering. Newer reservoirs now flank the structure.

About this time Basset & Coy in Wanganui began producing a small concrete-based panel of 38 mm thickness predominantly for house construction. It was supplied in a green state (partly cured) so that sawing, nailing and the external plaster finish could be carried out more readily. The board was secured to the exterior surfaces of timber framing to take the cement plaster finish. Styled 'Konka Board' this product was popular over the southern portion of the North Island and continued to be used for at least the next quarter century.

A bridge built in 1912 across the Ashley River near Rangiora is still in use although it has been modified. This structure consists of 24 continuous spans of 12.5 m, each with haunched T-beams carrying the deck on piers with piled foundations. Reinforced concrete was used throughout. The piers are unusual in having three hexagonal openings in each. The beams are not strictly haunched in the normal manner but are gently curved throughout their length to add some grace. In 1978 the solid balustrades were removed, a new kerb poured and splayed pipe handrails fitted to increase the effective width.

AUCKLAND TECHNICAL COLLEGE, WELLESLEY STREET, AUCKLAND.
This is an early use of concrete in an educational building dating from 1912. The architect was J. Mitchell. Good proportion and generous fenestration make this a pleasing building.

BUILDING, SWAN LANE (OFF CUBA STREET), WELLINGTON.
Designed by J.M. Dawson, this 1912 commercial building still retains simple, fresh and pleasing facades. It enhances the streetscape with more conviction than many contemporary structures.

DALGETY'S WOOL STORE, BLENHEIM.
In 1912 Dalgety and Coy built this wool store on a sharply tapered corner section. The south-light (saw tooth) roof is hidden behind the sinuous parapets. The building no longer exists.

WAIHENGA BRIDGE, RUAMAHANGA RIVER, NEAR MARTINBOROUGH.
This bridge, built in 1912, was designed by G. Laing-Meason, consulting engineer of Wellington. The photograph shows the provision for expansion with a small gap in the paired piers.

Earlier wool stores erected for stock and station agencies had been of timber frame and weatherboards, galvanised iron, or brick. Dalgety and Coy built an interesting wool store in Blenheim in 1912. It was on a tapering section between two intersecting streets with an acute angle giving a very narrow corner treatment. The roof consisted of south-light (saw-tooth) trusses and the parapet curved to provide a sinuous flowing line. The arched windows at the apex echoed this curve. Serious cracking in the concrete walls led to a decision to demolish the building in the 1980s.

In the Wairarapa the Featherston County Council replaced the Waihenga Bridge near Martinborough with a new structure in 1912. A multi-span reinforced concrete bridge, it is still in use and was designed by the Wellington-based consulting engineer, Gilbert Laing-Meason. With a total length of 233 m, it has fourteen main spans of 12.1 m and eight approach spans of 6 m. Six of the approach spans have been filled in at some stage and the original square concrete post-and-rail handrails have been removed in favour of splayed metal pipes to provide greater width. An interesting feature is the pairing of piers in two places, being an early example of provision for expansion. The superstructure has three ribs designed as cantilevered girders and stiffened by transverse webs about 3 m apart. Pile driving began on 11 January 1911 so presumably it was designed in 1910.

By this time reinforced concrete design was practised by most, if not all, New Zealand architects, for they could readily appreciate its advantages over brick and masonry structures in urban areas. Its resistance to earthquakes made it a sounder investment for owners. One architect in particular who went his own way in rejecting the universal trend towards machine-made and factory-produced materials was James Walter Chapman-Taylor. He had been strongly influenced by the Arts and

WINDBREAK, MAKARA HILL, NEAR WELLINGTON.
Built in 1913 this wall replaced a rotted timber structure erected nineteen years earlier. It is recorded that completion of the concrete work was delayed by strong winds.

WALL, KARORI RECREATION GROUND, WELLINGTON.
In 1912 a concrete wall was built along the main frontage of what is now called Karori Park. The design was completed two-and-a-half years earlier.

SUBSTATION, WAIKINO.
Erected by the Waihi Gold Mining Company in 1915 this large substation was used to reduce the voltage from its powerhouse at Horahora on the Waikato River. It was broken down from 50,000 to 500 volts.

Crafts Movement in Britain and believed that materials should be handcrafted wherever possible and that they should have natural qualities. Accordingly he was quick to appreciate the virtues of concrete and in 1913 the house he designed for a New Plymouth merchant named Burgess, at 26 Standish Street, had a concrete paved floor 75 mm thick in 600 mm by 600 mm slabs, although the walls were brick. Later he designed a number of reinforced concrete houses but even in these he used door and window furniture in wrought iron of his own design and fabrication. Chapman-Taylor also had a great love of hand-hewn timber finishes.

In 1913 a large substation was built at Waikino on the Ohinemuri Goldfield for the terminus of the 50 kV line carrying electricity from the Horahora generating station on the Waikato River. This building consists of four gabled bays with a high single-storey and solid unfenestrated walls. One side has a lean-to. With the closure of the Waihi Gold Mining Company's operations in 1952 the building has been stripped of its electrical equipment and left to the elements. In its heyday it fed power to the huge 200-stamp Victoria Battery nearby — one of the world's largest — while the transmission line of some 77 km was reputed to be the longest in the world at that time. The substation reduced power from 50,000 volts to 11,000 and again down to 500 volts. It also transmitted power to another substation in Waihi. It is a relatively crude building when compared to the much larger Lake Coleridge powerhouse built two years later.

To illustrate more fully the extent to which concrete was used in this country I have included examples of small and more mundane structures. Two such are the walls at Karori Park in Wellington, and on Makara Hill Road, a short distance away. The Wellington district is renowned for its wind and so it shouldn't be surprising to find a windbreak on an exposed hill road.

In 1894 the Makara Road Board had erected a timber windbreak at the summit of the Makara Hill to reduce the force of the northerly gales on the road traffic. When the timber rotted, the Karori Borough Council constructed a much more effective barrier of concrete in the form of a 145-mm-thick wall with buttresses on the windward side. A breakdown of costs gives no mention of reinforcing steel but does state that a bag of cement was three shillings and tenpence. Work was completed in December 1913, having been 'delayed by strong winds'.

This same body under its engineer, G.W. Brigham, had a wall built to enclose Karori Recreation Ground, today known as Karori Park. An observation of this structure resulted in the following statement in the *New Zealand Free Lance* for 13 January 1912:

Now that compulsory training is underweigh [*sic*] I am glad to see the Defence authorities have at last found time to attend to the fortifying of our city. I was at Karori the other day, and noticed the beginning of a great fortress they are building there. It is evident from this they intend to make this city a veritable Port Arthur, capable of withstanding the combined armies of Europe. As far as a layman could judge, this fort is going to be an impregnable stronghold of immense size, but why it is being built at the bottom of the hill puzzles me. I do not know much about fortresses but would have thought this fort would have been better built on the top of the hill near the breakwind on the Makara Road so as to command the approach to the city from Makara. The only possible explanation I can think of is that the present Mayor of Karori, to make amends for the lack of patriotism in his predecessor, has given a part of the Karori Recreation Ground as a site for this fort. If this be so, I hope the Defence Minister will return the compliment by calling it Fort Cathie.

The drawings for the wall and gate were completed by August 1910 but it was not until funds were available that quotations could be obtained and the work did not start until January 1912.

One of the early sea walls and esplanades to be built of concrete is at St Clair beach in Dunedin. This was opened in 1913 after many years of arguments, reports and delays. It was designed principally as a bulwark against the erosion of the sandhills but served as a promenade for the beach users. The wall consists of slabs poured in relatively narrow sections and with the date of each scratched into the surface. Recently the barrier fence along the top has been modified to an all-timber stockyard version. One of the original flights of steps remains unaltered.

Along the East Coast and in Hawke's Bay and the Wairarapa it was quite usual for wool from the remoter coastal stations to be taken out in surf boats to waiting small steamers. In the Bay of Plenty it became the practice in several areas to use scows, which could come close inshore in good weather and load bales of wool from the beach, using their own derricks. At Te Kaha a concrete shed was built on the edge of the beach for the storage of wool sent down from Maungaroa station, some distance up the Kereu River. Transport was by a narrow rough road. This shed predates 1922 and it was probably built some time after 1910. It is a simple rectangular building with 225-mm-thick walls, presumably reinforced. The floor is concrete and there is a sliding door, and a window in the rear wall. The roof is timber trussed and clad in corrugated iron now completely corroded in front. When no wool was being stored a whaleboat was kept in the shed, as spasmodic whaling

AWANUI RADIO STATION MAST STAY BASES, NEAR KAITAIA. One of two radio stations set up for ships' distress calls in 1913, the Awanui station had a mast 120 m high. This had three stays secured in massive concrete anchor structures. For all the world these resemble giant bomb-proof sentry boxes.

was carried out at Te Kaha from the nineteenth century up to the 1930s. Interestingly, the farmhouse at Maungaroa station is built of concrete, presumably at about the same time as the wool store.

In 1913 two radio stations were established in New Zealand to provide a listening service for ships' distress calls, and for communication with lighthouses having radio, as well as with a number of off-shore islands. Morse code was used for all such radio contact. One station was at Awarua near Bluff and the northern one was at Awanui, sited in Wireless Road a few kilometres north of Kaitaia.

The Awanui station, built for the Australian Telefunken Company, consisted of a substantial concrete building for the operators and, nearby, a surprisingly large power building with 300-mm-thick walls for the 35 kW 50-cycle alternator driven

WOOL STORE, TE KAHA, BAY OF PLENTY.
This beach wool store was built sometime after 1910 to store wool before it was loaded into scows on the beach. When the store was empty a whaleboat was kept in the shed, whaling having been carried out from Te Kaha up to the 1930s.

by a 56 kW 4-cylinder Gardiner kerosene engine that used compressed air for starting. The staff lived in timber houses — one for the superintendent, another for the engineer, and a large one for the operators.

The radio mast, 120 m high, was triangular in section, supported by a large steel plate resting on three heavy glass insulators. There were three stays, each being secured into massive concrete structures with shaped recesses to take the strain at the appropriate angle. Each of these pillbox-like bases has a strange 'sentry-box' appendage for no obvious purpose. Indeed today, without the mast and stays, these blobs of concrete present a strange sculptural quality as they stand aloof in a farmer's paddock. (*To page 144*)

SMALL COURTVILLE, AUCKLAND.
This is the second of A. Sinclair O'Connor's apartment blocks and was built in 1914. On the corner of Waterloo Quadrant and Parliament Street, it is in a precinctual area. The large balconies are unusual for this period and are richly embellished.

Opposite: ST MATTHEW'S CHURCH, HASTINGS.
This large Anglican church consists of a major addition in concrete to the 1886 timber neo-Gothic building. F. de J. Clere used Perpendicular Gothic to give some form of continuity, while expressing the plasticity of concrete in the coffered ceilings.

KARORI ROCK LIGHTHOUSE, COOK STRAIT.
Sitting off Tongue Point on a rocky islet only 30 m by 6 m in area, this light is frequently lashed by gales. Construction began in 1914, but with all concrete being made on shore and taken to the site by small boat there were long delays, aggravated by the gales and high seas. It was completed in September 1915 and has always been an automatic light. (*Photo: Alexander Turnbull Library. Ref. C 6125.*)

Opposite: WATER TOWER, HAWERA.
Settlements in flat county need some vertical elements in their townscape. An impressive example is the tower erected in 1914 to improve the water pressure for firefighting. It was designed by the Borough Engineer, J.C. Cameron, and a consultant engineer, S.T. Silver.

SEA WALL, ST CLAIR BEACH, DUNEDIN.
Opened in 1913 as a protection against erosion of the sand dunes, this sea wall became a popular promenade. The original barrier fence has been replaced and some steps altered.

Opposite: ASHLEY RIVER BRIDGE, NEAR RANGIORA.
Built in 1912 this 300-m-long bridge of 24 spans has hexagonal openings in the piers. The gently curved beams show a refinement of the haunched beam. The original solid balustrade has been replaced to increase the effective width.

During the First World War a military detachment of 65 men was sent to guard the station with the wartime responsibility of maintaining links with British and Allied shipping in the Pacific Ocean. At the end of 1927 Wellington Radio took over the free service to mariners and the Awanui Radio station was closed. All equipment was removed three years later.

A small house was built in Dunedin in 1914 using a construction method called Orotonu concrete. This seems to have been essentially reinforced concrete, for today the building is still standing sound and well maintained. Sited at 662 Cumberland Street it has been swallowed, but not digested, by the University of Otago high-rise development.

F.de J. Clere, who was mentioned previously in connection with St Mary's Church at Karori in Wellington, also designed in 'Camerated Concrete' the small St Alban's Anglican Church at Eastbourne in 1912. In 1914–15 the much larger St Matthew's Anglican Church in Hastings had major additions in concrete carried out to his design. The earlier church, built in 1886, is of timber, the work of Cyril J. Mountfort of Christchurch, and some may consider the juxtaposition of timber and concrete to be at odds. In this instance, however, they complement each other and the interior is particularly harmonious. Clere's use of a gently curved coffered ceiling has considerable appeal. His work includes the chancel, transepts, vestries, a tower and a chapel. Both architects have used Gothic forms although treatment differs markedly. The exterior of the original timber portion was stuccoed to increase the compatibility of materials. After the 1931 Napier earthquake the tower was lowered by about 7 m.

Contemporary views of architects and engineers on concrete technology are reflected in a report in the December 1913 issue of *Progress*. In a paper read at the Wellington Philosophical Society, the harbour board engineer, James Marchbanks, stated:

> The Wellington Harbour Board has already executed a large amount of concrete work and the scarcity of timber will necessitate more in the future. Practically none but colonial cement has been used lately, although its use is not insisted on in the specifications, and in general it easily conforms to the requirements . . . The Clyde Quay wharf has proved very satisfactory in this material, and a few difficulties met with in construction can be obviated in future by allowing more room between the reinforcing rods for the concrete to flow. The piles 18 in. [450 mm] square reinforced with four 1 in. [25 mm] rods drove much better than those with eight 1 $^1/_2$ in. [37 mm] rods, and some of them received as many as 150 blows per foot from a three-ton monkey falling 6 feet [1.8 m] on to a hardwood dolly.

> Mr J. Campbell, Government Architect, admitted that concrete will surely displace brick in future in architectural works. Its drawback is the difficulty of using its great strength economically without violating public taste, which has been educated to demand aesthetic treatment and substantial appearance. A surface covering of marble or brick was only begging the question, but progress had been made by using an aggregate composed of stones of pleasing colour and rubbing off the cement on the surface till these were exposed. An undetermined danger lay in the question whether cement was not injurious to many building stones.

In the same issue Peter Ellis, writing on 'Excessive Re-inforcement of Concrete', stated:

> The cohesion of the constituent particles of concrete is one of its most valuable features; it is therefore very unwise to insert too much iron or steel in the body of the concrete because it cuts up the mass into small portions, thereby lessening its cohesion and destroying its compressive strength. Just enough and no more should be the rule for reinforcement, were it not that concrete lacks tensile strength it would be better probably without reinforcement altogether. Heavy thick reinforcement members should also be avoided, four $1/4$ in. [6 mm] bars are often better than one $1/2$ in. [12 mm] and these should only be inserted where the material is subject to tension, woven wire of fine mesh should never be used and as a rule round rods and thick plain wire is better than mesh work or punched plating, and suchlike round rods give the maximum sectional area of metal with the minimum of interference with the cohesive strength of the concrete.

An Auckland building signifying the beginning of finely designed and well-appointed apartments is Courtville. Built in 1914, it is well sited in Parliament Street near the corner of Waterloo Quadrant and facing the historic Supreme Court House (High Court of today), St Andrew's Church and former Government House — an area steeped in history. The originators of this scheme were William Walter Stanton and Ernest H. Potter, two businessmen who engaged the little-known architect A. Sinclair-O'Connor to design Radnor, a brick and concrete flats block that was demolished by the city council in 1977. Stanton and Potter built a three-storey block of self-contained flats, today referred to as Small Courtville. It is unusual for its time in having large balconies with gently curved soffits between bay windows. It is an intriguing building with a very individual character and shows Sinclair-O'Connor to have been an architect of sensitivity.

TRIUMPH DAIRY FACTORY, NGAERE, TARANAKI.
The cost of this 1914 building upset the shareholders of the Ngaire (*sic*) Cooperative Dairy Company. Nevertheless, reinforced concrete became the preferred material for dairy factories because of its fire and earthquake resistance and ease of washing down. This factory has twin curved parapets echoing the gently rolling landscape.

The town of Hawera boasts a fine reinforced concrete water tower completed in early 1914. Agitation by insurance companies resulted in a tower to provide adequate water pressure for fire fighting. The Borough Engineer, J.C. Cameron, was the general designer with the consultant, S.T. Silver, providing the structural calculations. It is of interest to note that the work was not done by contract. Day labour was used under the supervision of Assistant Engineer C.P. Cameron. Some 51 m in height, the tower consists of a cylinder stiffened by buttresses but changing to an octagonal form with a circular balcony above. The top of the tower supports the main tank, a cylinder containing 454,600 litres. A lower tank holds half this quantity.

The interior of the tower has a central column or shaft of reinforced concrete with radiating beams supporting the several floors. There is natural lighting from windows of varying sizes. The main entrance door and grilled windows with little balconies have pedimented hoods. The Italian Renaissance seems to have inspired much of the detail and there is a Baroque feeling in the elaborate curving steps at the base, emphasising the classical approach. The overall effect cannot fail to impress, and being attractively sited in a small park the tower has become a pleasant place for rest and recreation. The latter can be enjoyed when the tower is open (entry is free) by climbing to the lookout balcony. This gives a splendid all round view, especially of majestic Mount Taranaki. In the generally flat or rolling country of the South Taranaki countryside this tower has magnificent landmark qualities, being visible for many kilometres from all directions.

It is worth noting that when the Hawera water tower was almost complete a strong earthquake caused it to tilt some 0.75 m from the vertical. By using anchors and careful undermining on the low side, with the first tank filled with water to help compress the deliberately dampened clay underneath, it was possible to right the tower in seven days. The excavated clay was replaced by concrete with the result that the structure has withstood many subsequent shakes.

One of the early reinforced concrete dairy factories still standing in Taranaki is that at Ngaere. Known as the Triumph Dairy Factory, it was built in 1914 for the Ngaire (sic) Cooperative Dairy Company established in 1893. When the Triumph factory was built there was considerable disquiet among the shareholders who opposed the spending of £6,000 — no doubt a

RAWHITIROA ROAD BRIDGE, PATEA RIVER, TARANAKI.
Opened in 1914 this rural bridge has a narrowness typical of the period. During construction the pier was scoured, causing it to become suspended from the superstructure. It was duly repaired and the bridge completed for traffic. The designer was F. Basham, the husband of the renowned Aunt Daisy of early radio broadcasting.

very large sum at that time for a small concern. One of the attractions of using concrete was its greater resist-ance to fire and earthquakes. Furthermore, its hard and durable interior surfaces made for cleanliness and ease of washing down.

The Ngaere factory is sited in attractive gently rolling countryside with Mount Taranaki as a backdrop. It makes a fine architectural statement with the boldly curved twin parapets at the front. For a period it was an indoor cricket centre and it was encouraging to see such a worthwhile use for an otherwise redundant older building. It is now an engineering shop for agricultural machinery assembly.

As has been mentioned there was a very strong interest in reinforced concrete bridge design in Taranaki during the first two decades of this century. The Eltham County Engineer, Frederick Basham, husband of the renowned Aunt Daisy who spent many years in radio from the 1920s, designed the Rawhitiroa Road Bridge over the Patea River in 1913. His contract drawings show a bridge with two spans, each of 18.28 m, supported on a central diagonally braced pier. During construction severe scouring of the pier caused it to become suspended from the superstructure — it weighed over 100 tonnes! Repairs were made and the bridge was duly opened for traffic in 1914, but it still has a permanent bow in it. The balustrades are solid with several raking buttresses on the insides and between these are raised rectangular panels with splayed corners. Like so many early country bridges, it is very narrow with only 3.35 m between the wheel-guards. The original drawings show a cambered wearing surface of road metal above the concrete deck.

Reference has been made to the first all-concrete lighthouse on Burgess Island in the Mokohinau Islands. Several others followed, one such being Karori Rock light situated near Tongue Point east of Cape Terawhiti to warn shipping passing through Cook Strait. In 1909 the wreck of the steamer *Penguin*, with the loss of 75 lives, off Karori Stream mouth had aroused widespread and vociferous calls for a lighthouse, but there were lengthy delays while the question of the best site was argued. The problems of construction were immense. It was sited on a rocky ledge about 30 m by 6 m, frequently lashed by both southerly and northwesterly gales, and only a part being a metre or so above water. Work started on the

19.8-m concrete tower in 1914, with the preparation of a suitable level base. One of the difficulties was the lack of space, with only six men being able to work on the rock at one time. The main construction team set up a base on Terawhiti beach, living in tents. All concrete had to be mixed on shore and taken to the rock in a specially built boat. Furthermore, work was limited to calm weather and even then scaffolding, equipment and partly set concrete was washed away from time to time. Eventually the foundations were satisfactorily constructed and thereafter progress was somewhat easier. Work was completed in September 1915, with the light being exhibited on 20 October. Not surprisingly this has always been an automatic light with no resident keeper.

An early concrete bridge which still survives, although it has been superseded, is that spanning Manukau Harbour between Onehunga and Mangere. Its construction dates from 1915 and it is a 246-m multiple span structure with haunched beams supported on piers, each being founded on four piles. The abutments are of bluestone and the handrails consist of three metal pipes supported on concrete posts. Over the years spalling concrete and corroded steel have left the bridge in a parlous condition. This has been aggravated by settlement, leaving the deck in anything but an even plane.

A young country in terms of its built environment, New Zealand does not have many ruined structures; a few, however, can be described as dramatic. One is the gaunt skeleton of the former freezing works at Waipaoa, about 24 km from Gisborne on State Highway 2. It is of interest to note that, of four abandoned freezing works now in ruins, three are located in the Gisborne/East Coast region. In addition to Waipaoa, Hicks Bay and Tokomaru Bay have industrial monuments to the aspirations of local farmers in the early part of the twentieth century. The fourth is at Taihape.

On 5 June 1915 a meeting was held to form the Poverty Bay Farmers' Meat Company Limited. By 26 June a start had been made on the erection of the buildings at Waipaoa. These were substantial blocks, two of which were three storeys, in reinforced concrete, with all work designed and supervised by L.G. James, an architect and consulting engineer specialising in freezing works design and construction.

The initiative and impetus for this venture came from William Douglas Lysnar, a well-known farmer, businessman, mayor and member of parliament. The speed with which the various buildings were constructed is nothing short of remarkable, even by today's standards when fast track procedures are used on occasions. There was some urgency because of the war, and a penalty/bonus clause in the contract induced speed of construction and installation of equipment by 1 December 1915. The works were built so as to have a siding from the nearby railway. Lines were taken from this to serve some of the buildings which also had overhead tramways for convenient carriage of products. In spite of being regarded as the most modern and efficiently planned freezing works in the Dominion there was no chain system of slaughtering and dressing in use at that time. Each slaughterman carried out a variety of operations in sequence.

From the outset the Poverty Bay Farmers' Meat Company wished to develop a direct trade with United Kingdom ports, especially Bristol, and in 1919 purchased its own vessel, the *Admiral Codrington*. This move was to ensure its independence of shipping companies. Unfortunately for the company the ship was far too costly and further expense was incurred in fitting insulation in the holds, resulting in the loss of a killing season's transport. The freezing works were now losing money and in 1923 they were sold to Vestey's, the large British meat and shipping company. Operations continued for a while but in 1930–31 the Waipaoa works closed and the equipment was transferred to the Kaiti works of the Gisborne Sheepfarmers' Frozen Meat Company. Today the roofless, floorless and windowless carcases of the main buildings echo the sounds of bleating sheep in the surrounding paddocks. The most complete building is the attractive office and store — a single-storey concrete structure having refined concrete bracketed window heads.

Following hard on the heels of the freezing works at Waipaoa was the establishment of works just south of Taihape. Here again a company was formed of local farmers to provide their own killing and freezing business, no doubt with the aim of having some control over their produce. For some reason there is a dearth of information on these works even at the local level. It appears they came into operation in 1916 and ceased in 1925 with the company being wound up two years later.

Today the remaining intact building and ruins of another can be glimpsed in a paddock on the right of State Highway 1 when travelling north. The intact building consists of a single-storey rectangular concrete structure having paired gables abutting a two-storey wing, also gable roofed in corrugated iron. The ruined block a few metres away has only the walls remaining. Part was of two storeys and the remainder has walls pierced by large window openings. The roofless block has a large open interior devoid of columns. As with other concrete ruins the play of light and shade can produce interesting and sometimes dramatic visual effects.

One of Chapman-Taylor's earliest reinforced concrete houses, Whare Ra, was designed and built in 1915 for an elevated site in Havelock North. The client was Dr Robert Felkin who, with his wife Harriet, was closely involved in London with the Order of the Golden Dawn, created *c*. 1880 and based on the fourteenth-century teachings of Christian Rosenkruetz. Following a split in the Order, Dr Felkin founded his own Order, known as Stella Matutina, which harked back to 'Masonic Rosicrucianism'. The Felkin family came to New Zealand in 1912 to establish this Order in Havelock North. Chapman-Taylor was a member and would have been gratified to have Dr Felkin as a client. (*To page 152*)

FREEZING WORKS, TAIHAPE.
This was another enterprise by local farmers and came into operation in 1916. Also short-lived it closed in 1925. One of the buildings is now a ruin.

Opposite: FREEZING WORKS RUINS, WAIPAOA, POVERTY BAY.
These ruins reflect the aspirations of local sheepfarmers. They founded the Poverty Bay Farmers' Meat Company Limited in June 1915 and construction began three weeks later. The architect and engineer L.G. James supervised the project as an urgent wartime requirement. It was completed by 1 December 1915 — an amazing feat even by today's standards. The works eventually closed in 1931 after the company had been losing money.

WHARE RA, HAVELOCK NORTH.
Designed by W.J. Chapman-Taylor, Whare Ra was built in 1915 for Dr Robert Felkin. The attractive house reflects the ideals of the Arts and Crafts Movement, a characteristic of Chapman-Taylor's work.

The falling ground provided a large reinforced concrete basement for the members' meeting room. On a stage there still stands a small heptagonal 'sanctum' decorated with the symbols of the Order. The basement has haunched beams and a clear span with the exterior wall being buttressed on the outside. The main floor of the house, also of concrete, has the living room, where Chapman-Taylor's predilection for exposed hewn timber trusses can be seen, together with the individualistic detailing that is his hallmark.

Another of the more elevated houses in Havelock North is Tauroa, built in 1916 for Mason Chambers, a member of a well-known Hawke's Bay sheepfarming family. Two years earlier, the 1887 timber homestead on the same site had been destroyed by fire. Determined that a replacement home should be resistant to both fire and earthquake, Mason Chambers was pleased to have external walls of double brick with cavity and a structural frame of reinforced concrete. The first floor is also of concrete, as is the roof. The choice of materials and in particular the quality of craftsmanship and attention to detail make this an outstanding work by the Auckland architect William Henry Gummer. It anticipates the Modern Movement in New Zealand domestic architecture and has its main elevations modelled to produce a lively effect with their changing planes. These have been rendered in white cement plaster with a pebble dash finish.

An impressive cantilevered copper canopy defines the entrance. One enters by a circular hall to a narrow vestibule and thence to a larger circular hall. Here the dominant feature is the superbly crafted branched staircase lit by an elegant curving wall of clear-patterned leaded windows rising 6 m. This staircase reveals the beauty of concrete as a plastic material. Opening from the hall is an exquisite dining room with a semicircular wall having the architraves and skirtings moulded to the curve. There are two other fine rooms on this level. The large library is a truly beautiful space and like the smaller room

TAUROA, HAVELOCK NORTH.
The work of W.H. Gummer, this is a splendid essay in large residential design. It was built in 1916 using a reinforced concrete frame, first floor and roof. It is a forerunner of the Modern Movement in domestic work in this country and is noteworthy for the fine interior spaces. Detailing and workmanship are excellent.

has exposed concrete beams, main and secondary, and French doors opening to the garden. In plan the house is shaped rather like an L, with the ascender leaning back at 60 degrees to provide a sheltered court.

The upstairs level has a semicircular gallery leading to bedrooms and bathrooms in both wings. The latter are well appointed with the original high quality fittings still in use. Every element of this house has been brilliantly conceived and executed. It is a remarkable fact that such a standard was achieved during wartime when skilled craftsmen and materials would have been scarce. The wrought iron balusters of interweaving circles were fabricated on site using a hand-operated forge blower. Much of the fine joinery and woodwork of jarrah and kauri was also made on site.

St Peter's Cathedral, Hamilton.
Originally designed in 1914 by Warren and Blechynden as a brick Anglican parish church, St Peter's was later built in concrete, opening in late 1916. It became a cathedral in 1926 with the formation of the Waikato Diocese, and was not completed until 1933. Concrete surfaces are plastered on both faces.

Tauroa is a house of magic. It must surely have rivalled any other house of its time anywhere. Several years ago I visited the world-renowned Fallingwater at Mill Run, Pennsylvania, in the United States, built in 1937–39 to the design of the illustrious Frank Lloyd Wright for Edgar J. Kaufmann. It is dramatically sited to cantilever over a rocky waterfall in a forest. It is truly a wonderfully lyrical concept, but somehow the interior, for all its interest in exploitation of levels, did not captivate me as much as that of Tauroa.

In 1915 the first state hydro-electric generating station came into service at Lake Coleridge. The jetty and intake structure built out into the lake is a striking sight when viewed from the hillside above. The position of the actual intake, being submerged, can be identified by the vortex on the lake surface. Some distance away is the powerhouse, also of reinforced concrete. There were serious problems to be overcome in its construction because of the accumulation of loose shingle from the nearby Rakaia River. This is a quietly impressive building with a major extension completed in 1927 following the same architectural design. The interior is noteworthy for the structural system of portal frames and crane rail beams to support the heavy-duty travelling gantry crane. This system has been followed for all successive state hydro-electric powerhouses in New Zealand.

Powerhouse, Lake Coleridge.
Coming into service in 1915, this was the first state hydro-electric enterprise in New Zealand. The building is quietly impressive in a majestic setting. The photograph shows the later extension.

Not all designs in reinforced concrete have been prepared with conviction. In 1914 sketch plans based on a brick structure were commissioned for a new St Peter's Anglican Church in Hamilton. The architects were John Warren and J. Blechynden and their brief was for a building that proved to be in excess of the available funds. A tender based on a modified version of the brick All Saints' Church in Palmerston North had to be declined after the tenderer increased his price before signing the contract. New tenders were called in August 1915 for a design using ferro-concrete — an instruction reluctantly accepted by Warren. Although it was anticipated that only the chancel and a portion of the nave could be built for the money in hand, the nave was completed with a temporary west end and porch built in timber. Dedication took place on 12 December 1916 by Bishop Averill.

John Warren, in his lack of enthusiasm for reinforced concrete at this time, was probably in a minority among fellow architects. Possibly it was because this commission was for a church and he had a preference for the traditional materials of ecclesiastical building. His partner, Blechynden, was an engineer and presumably was in favour of concrete. When St Peter's was eventually completed its appearance was considerably improved and today it can be regarded as an interesting early example of a church in which a relatively new design method and material were used. The concrete surfaces, both inside and out, are plastered. With the formation of the Waikato Diocese in 1926 St Peter's became the cathedral under the diocese's first bishop, Cecil Arthur Cherrington.

Arch bridges of multiple span usually have uniform shapes and sizes for the arches. This is not so in the little railway bridge built in 1916–17 over Leaning Rock Creek in the Cromwell Gorge. Now covered by Lake Dunstan are five arches with the main span of 13.7 m being a parabolic arch flanked by segmental arches of 7.6 m. These are supported by the main arch, piers and the abutments. The solid concrete balustrade tended to give the side elevation an impression of being masonry — most unusual in a concrete bridge. It was not readily visible from the road except for a short time during the construction of the new high-level highway, when a temporary road downstream of the rail bridge gave a view of the side elevation. This structure is no longer visible. It was one of three reinforced concrete arch rail bridges in the gorge designed by J.E.L. Cull, the first designated design engineer in the Public Works Department.

A Masterton architect, F.C. Daniell, moved to Hamilton in 1908, spending about 27 years there before returning to his home town. During his northern sojourn he designed many buildings in reinforced concrete, having partnerships with Andrews (c. 1912), Gray (1914–17), and then with someone named Lush, for a period. His early work in Hamilton made use of 'Camerated Concrete'. The New Zealand Cooperative Dairy Company engaged him to design several dairy factories, a good example of this work being the Matangi Dairy Factory built in 1917. It is of column and beam construction with concrete panel walls and timber-trussed roofs — a straightforward and pleasing building still in use, although major buildings have been added. A versatile man, Daniell was responsible for a small concrete bridge in 1911 at Mangahoe for the Waipa County Council.

No doubt F.C. Daniell is typical of a number of lesser known architects of the first two decades of the twentieth century who worked quietly in the design of reinforced concrete without the services of a structural engineer. Having to confront the architectural problems of handling a construction medium with no proven aesthetic finish often produced some pedestrian results, but their approach was mainly honest. The more successful development of finished surfaces involving texture, off-the-form, fairface and suchlike, apart from traditional plaster, came after the Second World War.

Various materials have been used in the construction of lighthouses in this country, including timber, cast iron, stone and concrete. One that is a unique blend of both stone and concrete, although not originally intended as such, is Dog Island Lighthouse. Situated on a small, bare, windswept, low-lying island in Foveaux Strait it is New Zealand's tallest lighthouse, safeguarding shipping using the Port of Bluff and the strait. The designer was the highly competent James Melville Balfour,

LIGHTHOUSE, DOG ISLAND.
On a small, bare and low-lying island this 36-m-high lighthouse was built in stone in 1865. It was designed by J.M. Balfour. Early decay of the mortar and cracks necessitated strengthening with external vertical timbers. In 1916 major repairs were carried out using a poured concrete sleeve on the outside and an inner concrete skin up to the fourth floor level. (*Photo: C.D. Kerr*)

CHIMNEY, TANE HEMP COMPANY'S FLAXMILL, OPIKI.
This mill was one of many in the Manawatu and was close to the Rangitane bridge the company built. The concrete chimney is a mute and solitary reminder of the noise once made by the steam-driven stripping and scutching machines.

an uncle of Robert Louis Stevenson, who had trained in lighthouse design and construction under D. & T. Stevenson, the renowned lighthouse builders of Britain. His design was for a stone tower 36 m high, which was completed in 1865 using stone quarried on the island. However, as the mortar soon decayed and cracks had developed by 1871, strengthening was warranted. John Blackett, Public Works Department Acting Engineer-in-Chief and Marine Engineer, used heavy timbers vertically which were fastened by iron bands to overcome the problem. In 1916 the tower was once more requiring attention. Two years earlier it had been struck by lightning which may have aggravated or even caused the weakening. The original weak mortar had been replaced by Portland cement mortar early in its life, but this time the more drastic measure was taken of pouring a reinforced concrete sleeve on the exterior, giving it a greater diameter. This addition was 609 mm thick, and at the same time an inner skin of 150 mm was added up to the fourth-floor level. Dog Island Lighthouse is still in use.

In 1917 the government of the day built the concrete Manorburn Dam in a remote highland part of Central Otago between the Knobby Range and Rough Ridge. This was a major work in carrying out a policy of irrigation for the extremely low rainfall areas. The dam is a concrete arch structure with gravity abutments on foundations of the ubiquitous Central Otago schist. With a height of 33 m it impounds the biggest volume of water of the several early dams in the region. Like the Brook Street Dam in Nelson it suffers from poor sand quality in the concrete mix and has had problems of water retention. Wire mesh reinforcement was used close to the face of the dam. One of the abutments required major reconstruction in an attempt to overcome the deficiency of extensive leaking. The Manorburn Dam, although small when compared with present-day hydro-electric dams, nevertheless presents an impressive sight from downstream where schist, tussock and tumbling water provide visual delight. For many years this dam has been a popular venue for ice skating.

The earliest and oldest extant reinforced concrete bowstring arch bridge in New Zealand, begun in 1915, spans the Opawa River in Blenheim. Of eight spans of 21.33 m, this bridge employs the principle of rather heavy concrete through arches, in which the deck passes through the arch and is supported by vertical hangers from the top chord. These hangers are tension members and in later bowstring arches became quite slender. Here, however, they are noticeably solid and, moreover, they have chunky diagonal braces. These arches are rather flat — again in strong contrast to the higher, more graceful flowing arches of some of the bridges designed in the 1930s. The Opawa Bridge, opened in December 1917, was the work of the Public Works Department and is still carrying the much heavier loads of State Highway 1 traffic. The period 1930–40 saw the bowstring arch achieve a measure of popularity in some countries such as Britain and France. In New Zealand PWD engineers and a few consultants designed a number which are still in use. In the United States they are termed rainbow arch bridges.

In 1917–18 a 145.4-m suspension bridge was erected over the Manawatu River by the Tane Hemp Company Limited at a place then known as Rangitane. This bridge, built to facilitate access to the company's new flaxmill, has twin reinforced concrete tapered towers, 14.6 m high, with three lateral braces in the same material. The tower foundations go down about 3 m. The designer and contractor was Joseph Dawson of Pahiatua. Interestingly, the sixteen wire ropes of 34.2 mm diameter had seen earlier service at a Waihi goldmine. When the Tane flaxmill and others in the vicinity closed in 1921 a local

CHURCH, MANAKAU, MANAWATU.
This former Methodist church of 1918 shows an early use of concrete blocks. The walls have external buttresses and the blocks are rusticated to resemble stone. It has been converted to a residence.

landowner, Hugh Akers, took over the bridge and called it the Opiki Bridge after his property. He charged users a toll to cover maintenance costs and this continued until 1969, being the only privately owned toll bridge in New Zealand. When it closed following the opening of a new bridge, the Manawatu Catchment Board, as the new owners, removed the deck and stiffening truss for safety reasons, leaving the loose hanging vertical cables to give a weird and forlorn appearance.

Standing aloof near the south end of the bridge is the sentinel-like concrete chimney of the former Tane flaxmill owned by the Tane Hemp Company. Built in 1916 it was the smokestack for the steam engine that provided power for driving the stripping and scutching machines. Both the bridge remains and the chimney have been classified by the New Zealand Historic Places Trust as industrial monuments meriting preservation and as reminders of a once widespread flaxmilling industry.

An early use of concrete blocks, or at least an unusual one for churches, can be seen in the former Methodist church at Manakau in the Manawatu district. It was built in 1918 as a simple, gabled, rectangular structure of concrete and concrete block walls with external buttresses. The roof has a squat louvred tower surmounted by a diminutive spire. An ecclesiastical appearance is given by the pointed arches of the door and stained glass windows. The concrete blocks are rusticated to simulate a stone texture and for a period were largely festooned in creeper. With no off-street parking available on its small slightly elevated site, an acute problem developed for church members in parking on the very busy State Highway 1. A fatal accident to one of them while entering a car hastened the decision to close the church and amalgamate with the Anglican congregation. Bought privately after its closure in the early 1970s, it subsequently changed hands and the new owners have converted it to a residence.

The Opawa River Bridge at Blenheim was followed in early 1918 by the Westshore Bridge at Napier. It has a single span bowstring arch of 17.37 m with 21 spans of 15.2-m concrete and steel plate girders and also five of 6 m. The bridge also carries the Napier–Gisborne railway and has been modified and repiled as a result of damage in the 1931 Hawke's Bay earthquake. It has been much less used since the main highway was re-routed.

This bridge makes an interesting contrast with the reinforced concrete arch structure over the Shotover River at Arthur's Point near Queenstown. It was designed by Frederick William Furkert of the Public Works Department who was to become

EDITH CAVELL BRIDGE, ARTHURS POINT, NEAR QUEENSTOWN.
Probably inspired by Grafton Bridge in Auckland, the designer, F.W. Furkert of the Public Works Department, produced a dramatic structure 30.48 m above the Shotover River. The parabolic arch seems entirely appropriate in this rugged site. The open spandrel supporting the deck retains a sense of lightness which is carried through in the open balustrade.

engineer-in-chief of that robust department. He was inspired most probably by the magnificent Grafton Bridge in Auckland but he had a more dramatic landscape in which to fit his concept. The choice of the concrete two-pin parabolic arch made this only the second parabolic concrete arch bridge to be built in New Zealand — furthermore it was constructed in record time. Opened in 1919 it has a span of 30.48 m with a height above water level of 27.4 m. The roadway is 4.26 m wide, which is narrow by today's standards. This bridge follows the Grafton Bridge in having the deck immediately above the crown of the open spandrel arch, but it has horizontal ties to the vertical struts. The abutments are faced with schist — the

CORNER COURTVILLE, AUCKLAND. Sited on the corner of Waterloo Quadrant and Parliament Street this wedge-shaped building has fifteen self-contained flats. The architect, A. Sinclair-O'Connor, showed considerable skill in the design and interior detailing of this dignified building.

prevailing stone of Central Otago. With the very considerable impact of tourism in the Queenstown district this is now a favourite stopping place from which to watch the tourist-laden jet boats negotiating the turbulent waters in the gorge below.

There is a story about an old miner, Jack Clark, who lived nearby in his little sod hut. He was an ardent admirer of Edith Cavell, the British nurse shot by the Germans in the First World War for helping British and French soldiers escape from Brussels to the Dutch border. Wishing to have the bridge at Arthur's Point named after her, he first painted 'To Cavell Bridge' in large red letters on the rocky bank a short distance away and asked the county council for it to be officially named after his heroine. This request was not acceded to and later he went so far as to paint the name on the side of the bridge. Eventually people came to accept Jack Clark's wish, for today it is well known as Edith Cavell Bridge instead of the earlier name of Upper Shotover.

The large flour mill at Gore consists mainly of three contiguous buildings and is said to be the largest cereal producing plant in the Southern Hemisphere. The original building was erected in 1878 for Richardson, Greer and Company but this was replaced in 1919 by a reinforced concrete block of five floors. It is flanked on one side by a smaller three-storey building and on the other there is a three-level concrete silo built in 1912. At the time the new block was being built the silo roof was altered to form a penthouse. It is a very functional looking structure with its blank walls now decorated by the firm's long-standing porridge character, Sergeant Dan. There is a large Scotch thistle on the end wall. The newer (1919) block is of reinforced concrete beam and column construction with large windows having shallow concrete spandrels. Early in its history the mill was acquired by Fleming, Henderson and Coy and it is generally known today as Fleming's Mill, notwithstanding its ownership by a national group of companies.

A rather conventional reinforced concrete arch bridge was designed in 1919 by Jones and Adams, consulting engineers of Auckland, for the Horotiu crossing of the Waikato River north of Hamilton. Opened in 1921 it has a single three-hinged arch of 38.4 m with the deck passing through the apex. The balustrade has minimum visual impact, thus giving emphasis to the main structural members.

In 1919 the corner site of Parliament Street and Waterloo Quadrant in Auckland was taken up with the five-storey wedge-shaped Corner Courtville, having fifteen self-contained flats. This was a construction venture by the energetic James Fletcher of Dunedin, founder of the highly successful Fletcher Construction Company. He was knighted in 1946 for his services to building. Corner Courtville is a splendid building which treats the corner site with dignity and aplomb. It has bay windows and deep-set balconies to all floors. The roof line is defined by wide overhanging eaves and the setback on the corner is surmounted by a small domed tower. Courtville illustrates the competence of Sinclair-O'Connor as a designer. The interiors of both Corner and Small Courtville are of a very high standard in their appointments and detailing.

The Auckland architect R.W. de Montalk was mentioned in regard to the Hotel Cargen of 1912. About 1919 he produced a design for an interesting concrete house in Wellington. He was an early advocate of reinforced concrete and his son believes he designed houses in this medium from as early as 1902 in Auckland, but I have not been able to verify this. Certainly in the Grosvenor Terrace house in Wadestown he went further in his use of precast units such as roof tiles. Columns, beams and even roof framing members are of reinforced concrete. The house of one-and-a-half storeys with a small basement must have had very considerable delays because it was not completed until 1925.

One of several splendid Hawke's Bay houses designed by William Henry Gummer of Auckland is Craggy Range, on farm land southeast of Havelock North overlooking the Tukituki River. Dating from 1919, it was built for Ivan van Asch and was not completed in its entirety for many years. Ivan's brother, Piet, founded the pioneer Aerial Mapping Company. The house is of reinforced concrete frame, first floor and roof, but with double-skin brick cavity walls. This system survived the disastrous Hawke's Bay earthquake without damage. The house, now known as Belmount, anticipates the Modern Movement in New Zealand and seems to have been ahead of the early European examples. It is beautifully composed in its elevations, expressing horizontality and calm.

FLOUR MILL, GORE.
In 1919 this five-storey building replaced the original one of 1878. Reinforced concrete beam and column construction gives generous windows for good light. The adjoining block was erected in 1912 as a silo and depicts Sergeant Dan. This character was pictured on the popular Creamota porridge packets, the firm having concentrated on oat cereals for many years.

1 GROSVENOR TERRACE, WELLINGTON.
The architect R.W. de Montalk was interested in concrete construction from the turn of the century. This house was designed *c.* 1919 and was innovative in having a total concrete concept. Walls, columns, beams, rafters and roof tiles are all concrete.

INTAKE AND JETTY, LAKE COLERIDGE POWER STATION.
The submerged intake can be identified by the vortex in the lake waters beyond the jetty.

CRAGGY RANGE, NEAR HAVELOCK NORTH.
Designed in 1919 this is one of several very fine houses of W.H. Gummer to be built in this district. He used reinforced concrete for frame, first floor and roof, with brick walls. This house, now called Belmount, is splendidly composed in its elevations, having horizontality and an affinity with its site. It anticipates the work of European architects.

Above: RANGITANE BRIDGE, MANAWATU RIVER, OPIKI.
Built for the Tane Hemp Company in 1917–18 this bridge provided access to the new flaxmill. The concrete towers, 14.6 m high, support the cables from which the hangers are suspended. The deck was removed when the bridge was replaced. The designer and builder was Joseph Dawson. After the mill closure this bridge became privately owned and tolls were charged.

Top left: MANORBURN DAM, CENTRAL OTAGO.
This remote dam is a concrete arch with gravity abutments founded on schist rock. It was built in 1917 by the Public Works Department for irrigation in this very dry region. It is also used for ice skating.

Bottom left: LEANING ROCK CREEK BRIDGE, CROMWELL GORGE.
Now covered by Lake Dunstan this rather odd-looking arch bridge was designed by J.E.L. Cull of the Public Works Department. It was completed in 1916 for the Otago Central Railway.

Opposite: OPAWA RIVER BRIDGE, BLENHEIM.
This is the oldest concrete bowstring arch bridge in New Zealand. Designed by the Public Works Department and completed in 1917 it was a pioneer example of this type. By comparison with later bridges of this form it shows rather cumbersome proportions. It is still in use on State Highway 1 and carrying much heavier loadings than were originally intended.

Chapter Nine
1920–29

The third decade shows a much greater acceptance of reinforced concrete as the all-round construction medium. Structural steel suitably encased in concrete was used to a limited extent and in some taller buildings. It was also used in some bridge designs, principally as plate girders but also in the form of trusses. Timber was still quite common for bridging, either as simple beams or as trusses, the latter often in composite form using some steel, but reinforced concrete was gaining ground steadily. The fact that it neither rotted nor rusted and was resistant to both fire and earthquake made it a sounder investment. Moreover, building and civil engineering contractors were coming to terms with its construction methods and requirements.

Dating from about 1920 is a stark concrete ruin in a paddock near Kotuku on the West Coast. It is a timber drying kiln at the site of the former Jack Brothers' timber mill, which originated in 1902 and was one of the largest mills in the district in the earlier part of the century. Kiln drying of timber goes back to the 1870s, although I am not sure when such treatment began in New Zealand. It is necessary to reduce the moisture content of the newly cut timber from around 50 percent to about 18–20 percent to avoid undue shrinkage and movement when used in building. It was common in my boyhood days to see timber yards with timber stacked like a wigwam but in long rows for air drying — a process that required many months.

The kiln at Kotuku is probably a fairly early example. It consists of a 21-m-long box culvert in form with two square flues at one end. One on each side served the boiler, which is no longer in evidence. This provided heat to a space about 1.5 m deep below the floor. The exterior has piers at 3-m centres with the concrete walls giving every indication of formwork made by bush carpenters. Aloof in a paddock, it is a strange and baffling structure to the uninitiated.

During this period F.de J. Clere was still practising in Wellington, designing churches in his capacity as Anglican Diocesan Architect, as well as secular buildings. Having already established that reinforced concrete was a suitable material, he designed the little church of St Matthias at Makara. This was a replacement for a timber church built in 1867. Very much a country church in its setting, it rests happily on a slightly elevated site in a spacious churchyard surrounded by the graves of worshippers whose names go back to pioneering days.

The Church of St Matthias opened for worship on 21 August 1921. It is simple in form, with neither transepts nor apse, but there is a square tower giving it a Norman appearance. The fine interior has a timber ceiling supported by timber trusses and the walls are plastered, as is the exterior but in pebble dash (roughcast) finish. Although Clere used Gothic elements, such as the pointed arch, in his churches, some of these small concrete essays have a Romanesque feeling with the solid walls being pierced by relatively small windows. In Lyall Bay, Wellington, St Jude's Church is such a one.

Previous page: TIMBER KILN, KOTUKU, WEST COAST.
Probably dating from *c.* 1920 this unusual-looking structure was built to dry freshly sawn timber from the former mill on the site. A boiler supplied heat through a space 1.5 m deep under the floor. The two flues are dominant at one end.

    The Luttrell brothers, already referred to in chapter seven, forged a considerable reputation for their design and construction of racecourse stands and associated structures. At Christchurch the Canterbury Jockey Club required a new main grandstand to replace one destroyed by fire in 1919 at the Riccarton racecourse. Construction began in 1920 and was completed in 1923. Of reinforced concrete, it has four storeys at the rear with access by prominent ramps. The two levels of seating have protection from the partly cantilevered steel trusses. On the end elevations the concrete framing is boldly expressed and capped by gables that are echoed in the gablets of the main roof. The design relies not on applied ornament for interest but on the functional forms dynamically portrayed.

    The Wellington Racing Club had engaged Sidney Luttrell in 1916 to advise on an extensive building programme at the Trentham racecourse. In 1919 he produced contract drawings for three stands — stewards', members' and public. The latter grandstand is a large structure of reinforced concrete, and the same material is used for the dominant expression of ramps and stairs for the main access at the rear.

    Difficulty in obtaining materials delayed construction but Sidney overcame the cement shortage by buying into the Golden Bay Cement Company. Very soon afterwards he was a director, and in 1923 managing director, thus ensuring a guaranteed supply of cement and completion of the grandstand in 1925. The tiered seating of concrete is supported on substantial steel trusses and the ceilings in this area are of pressed metal. The balustraded front of the stand has attractive slates. Concrete surfaces are lined out to break up the large scale; however, much of this is now covered in creeper, adding to the structure's appeal. Every aspect of the Trentham grandstand has been most carefully resolved to produce a superb piece of architecture that serves its function admirably.

    Occasionally substantial industrial structures were built in remote and sparsely populated areas. One example was the freezing works building at Hicks Bay. The Hicks Bay Farmers' Meat Company was convinced that the construction of a wharf nearby would allow overseas vessels in the roadstead to be serviced by lighters. At this time roading was so primitive in this district that all material for both wharf and freezing works had to be brought in by the coastal ships of Richardson & Company. The works were completed and operational in 1921, being built of reinforced concrete throughout with timber-trussed roofs of corrugated galvanised iron. The first floor concrete slab was supported by a series of finely proportioned columns with the beams taking the form of shallow arches — a surprisingly elegant design for a freezing works.

    As with many early industrial enterprises parochial enthusiasm overran prudence and the project foundered in a very short time. In a year the wharf had collapsed through dry rot and heavy seas. In 1925 a new wharf was built but a year later

St Matthias' Church, Makara, near Wellington. Replacing an earlier timber building, this delightful church was designed by the Diocesan Architect, F. de J. Clere, and opened for worship in 1921. Attractively sited on rising ground with a graveyard, it has a Norman-like tower. The fine interior is simple but most effective.

the freezing works closed as an uneconomic venture. In such a lightly populated district no alternative use was forthcoming and the building became a partial ruin. Eventually a portion was made habitable as a residence and the ground floor used for storage.

Clere certainly advanced his design concepts for concrete ecclesiastical buildings when in 1919 he called tenders for the large Catholic parish church of St Mary of the Angels in Boulcott Street, Wellington. An earlier timber church, the second on the site, had been badly damaged by fire in 1918, necessitating a new building. The successful contractor, H.E. Manning, seriously hampered by post-wartime shortages, relinquished the contract early in 1920. Thereupon the parish priest, Father Mahoney, with the assistance of a friend, Martin Malony, became clerk of works, employing day labour to complete the work under Clere's architectural supervision. With money dependent upon the Sunday offerings there were very considerable difficulties. It says much for their faith and perseverance that this large church was completed, with the opening for worship taking place on 26 March 1922.

St Mary's is a striking architectural symbol in the Wellington inner-city streetscape. Sited at an angle to the rising street, it has a noble west front

Main stand, Riccarton Racecourse, Christchurch. The firm of S. and A. Luttrell of Christchurch designed and built several grandstands for racing clubs. This impressive concrete stand was completed in 1923. It has four storeys with the ends showing generous windows.

(liturgically speaking) with the main entrance reached by solidly balustraded, asymmetrically placed steps to a balcony. Very much an essay in the Gothic style, it has French overtones. However, it is believed that Clere was inspired by the Cathedral Church of St Michael and St Gudule in Brussels as he had a design for the Anglican cathedral, never built, showing this influence. The reinforced concrete structure is faced in part with a mellow brick and the exposed concrete has a grey cement plaster finish. After strengthening and refurbishment in the 1980s the exterior has been painted a cooler and probably less successful grey hue. St Mary's can boast some early precast work with such elements as the large rose window in the west front. Indeed, the crafting of traditional Gothic decorative elements exploits the plasticity of concrete.

The splendid interior is notable for the use of remarkably slender reinforced concrete columns and pointed arches supporting the timber roof members and ceiling. There are also Gothic arches, which extend from the aisles around the apse with its imposing High Altar. These are repeated in the clerestorey windows. It has been claimed that this use of reinforced concrete ribs for the roof was the first such ecclesiastical example in the world. It was certainly a bold innovation for its time. St Mary's is a superb piece of architecture. A highly successful public appeal for money for strengthening and deferred maintenance has enabled a start to be made on this. The original slate roof has been replaced in copper in the traditional sheet form with the intention of eliminating the leaks caused by broken slates. St Mary of the Angels ranks as one of the finest churches in New Zealand and marks the zenith of Clere's work.

A rarity in New Zealand is the Gentle Annie Creek Bridge about 20 km from Cromwell. A tributary of the Kawarau River it is crossed by a reinforced concrete bridge using a modified Pratt truss of 18.28 m with solid concrete end panels. The smaller span is a 12.19-m girder and the pier consists of a paired horizontally braced reinforced concrete bent. Concrete trusses are most unusual. The bridge was designed by the Public Works Department and completed in 1922.

Several memorial bridges have been erected over the years, some of them in rural areas. One is the Stewart Creek Bridge at Island Block between Raes Junction and Ettrick. It was built in 1922 in memory of Robert and Isabella Stewart by their children and gifted to the Tuapeka County Council. This is a small bridge with large square posts capped with spheres terminating the wrought iron balustrades. One of the posts bears the memorial tablet.

Another small but intriguing bridge in a country setting can be seen at Kaiparoro in the Wairarapa near the Mount Bruce Wildlife Centre. This Anzac memorial bridge, built in 1922, was designed by A. Falkner and the Mauriceville County Engineer was W.A. Miller. The bridge has a single span, being a rather flat arch with solid balustrades. The eastern one has a centrally placed circular motif protruding above the handrail level. It bears an inscription above the names of the fallen from

GENTLE ANNIE CREEK BRIDGE, KAWARAU GORGE, CENTRAL OTAGO.
This bridge is unusual in having a reinforced concrete truss in addition to the conventional girder span. The Public Works Department designed the bridge, which opened in 1922.

the district who served overseas. Something of a curiosity, this bridge lacks grace and some years ago was painted in stark white with lettering picked out in vivid blue. If it seemed dull in its weathered grey state it now appears rather garish. It has been superseded by a modern structure alongside.

Having lived in Auckland for several years in an early concrete house, I soon became aware of the problem of condensation on internal walls. This is very noticeable in areas prone to high humidity. In 1922 an American, Oswell C. Hering, wrote a book entitled *Concrete and Stucco Houses* in which he advocated the use of furring to overcome condensation. This consists of wooden strips nailed to the concrete wall to take laths for plaster. The method leaves an air space between concrete and plaster. In his book Hering was generally critical of the standard of concrete blocks being produced throughout the

United States. The 'dry' process had been used for a number of years and he stated that too dry a mix caused low strength and crumbling. A new, properly made, 'real' concrete hollow tile block had become available and showed much promise. He deprecated the 'art' or 'rockfaced' block as being spurious in appearance, for in his view a concrete block should stand on its own merits.

Concrete as a road material originated in the United States. About 1893 a concrete pavement was laid in Bellefontaine, Ohio, with a 100-mm base having a 50-mm wearing surface. Such pavements were advocated for residential streets. At the end of 1902 a total of 9.6 km of concrete was used in highway roading. At this time there were 127,731 cars in that country; by 1919 there were almost 8,000,000 and the length of concrete roads was 18,240 km. Before the First World War the work on concrete roading was largely experimental and there were some failures, but the demands of wartime industry and heavier truck traffic resulted in better standards being achieved.

One of the first countries outside the United States to take up concrete as a road material was Australia in 1914, with New Zealand following in the next year. In Auckland this country's first concrete road was constructed under the direction of the City Engineer, Walter E. Bush, and was considered highly successful. Even before this, concrete foundations had been used for all the city's paved streets. Early examples of completely concrete roads were Little Queen Street, Durham Street and Park Road.

In 1916 an Auckland publication, *The Roading of the Future: A Few Facts about Concrete Roads* produced by Wilson's Portland Cement Company Ltd, advocated that the construction of concrete paving for roads in New Zealand should be based on observations of the effectiveness and economics of concrete roads in the United States. Presumably it had some effect on local authorities and their engineers because there was a period in the 1920s when concrete roads were laid down in several districts.

After the First World War the Dunedin City Council adopted a policy of upgrading its streets, especially those carrying heavy traffic, by laying a foundation of 178 mm of concrete and then applying 50 mm of Trinidad asphalt. At this time the City Engineer was W.D.K. McCurdie.

Even the small borough of Thames could see the advantages of concrete paving and the Consulting Engineer, E.F. Adams, produced contract documents in November 1923 for concrete in Pollen Street, the main thoroughfare, for a length of 1.3 km. This still remains.

Concrete pavements were constructed for the Great North Road and the Great South Road as the approaches to

Town Library, Puhoi.
This diminutive library must be one of the smallest public buildings in the land. Pleasantly sited on the bank of the Puhoi River it makes a modest contribution to the historic village founded in 1863.

Staff houses, Portland.
These staff houses were built for Wilson's (NZ) Portland Cement Ltd after it was formed in 1918. Some earlier houses have been demolished and these double units date from the early 1920s.

Auckland. Likewise in the 1920s there was a concrete road between Napier and Hastings and there were also other districts with concrete paved roads.

A very simple concrete building of diminutive size can be seen at Puhoi, 56 km north of Auckland. Built in 1923 it is pleasantly sited on the riverbank, making a modest contribution to the historic village which began in 1863 when a group of immigrants from Bohemia settled here. The other buildings including the Church of SS Peter and Paul, convent school, presbytery and hotel are of timber, but the concrete library is in harmony with its neighbours with its unpretentious design and white plaster finish.

In 1924 the do-it-yourself ability of New Zealanders was demonstrated when James E. Davis built himself a house in Albert Street, Palmerston North. He made his own panels using Portland cement, crushed pumice and pig's hair, the dimensions being approximately 800 mm by 500 mm by 50 mm. Presumably this had a similarity to 'Konka Board', for his daughter remembers a visitor checking to ensure that no patent was being infringed. The house is still standing and, with some modifications, is used as a medical centre.

As evidence of faith in concrete construction it is only to be expected that a cement manufacturing company would build its staff houses of concrete. This is so at Portland — the settlement that evolved following the formation in 1918 of Wilson's (NZ) Portland Cement Limited (see page 89). The new company decided to erect several staff houses in concrete, and these can be seen today as a row of double units on sloping ground below the road. Each unit is single-storeyed and has a basement. With their tiled roofs they form a pleasant group dating from the early twenties. The works manager's house has a different design and a fine view over Whangarei Harbour; it was awaiting refurbishment when I saw it in early 1988. There are two derelict concrete houses nearby. Several other houses constructed earlier with flat concrete roofs have been demolished.

A church was built of concrete and in 1921 a wooden community hall, now stuccoed, was brought over from Limestone Island. Alongside this a concrete library was built in 1924. Today the village presents an attractive environment. Its name, of course, derives from the Portland cement produced there.

Several years ago while travelling in Washington State in the USA, my wife and I came upon a small town called Concrete. Sited in the Upper Skagit Valley it had once been known as Cement City after a cement plant was opened late last century. In 1908 another company began production not far away and the two settlements merged to become Concrete, with the inevitable takeover of one of the cement companies. There was a considerable boost for the product between 1918 and

1961 when Seattle City Light built three large dams and powerhouses on the Skagit River in the Cascades. The cement plant closed in 1968 but the town survives today. Its largest and more recent building is Concrete High School, part of which straddles a street. Apparently, when the town was being established, there was a proviso in the sale of lots that buildings must be of concrete only.

The firm of consulting engineers Jones and Adams, which later developed into KRTA Ltd, was responsible for a goodly number of early reinforced concrete bridges in the Auckland and Waikato regions during the 1920s and thirties. Fairly certain to have come from this office is the design for the oldest bridge in Manukau City. Built in 1923 it is the Flatbush School Road No. 1 Bridge, which is virtually in its original condition. It is of single span with the deck supported on three haunched beams. The sides of the abutments have unusual buttresses. The balust-rades, so often solid and heavy on early concrete bridges, have three pipe rails supported by square concrete posts with pyra-midal caps — pleasing and functional for the period.

FLATBUSH SCHOOL ROAD NO. 1 BRIDGE, MANUKAU CITY. This small bridge was built in 1923 using three haunched beams. It was designed by Jones and Adams of Auckland. The illustration shows the open balustrade — a practical and pleasing design.

From the time of successful refrigeration on overseas vessels in 1882 until the first decade of the twentieth century there was a spate of dairy factory building, mostly in timber. The replacements for these and for new ventures saw a change to brick or concrete. Understandably concrete was a superior material for this type of building as it enabled the interior spaces to be thoroughly and hygienically cleaned down. Although many early factories are now abandoned or lie derelict, some have been put to new uses.

A good example of this recycling is the dairy factory at Ngatea built probably in the early 1920s. When the New Zealand Cooperative Dairy Company Limited took over in 1927 it already had a history of mechanical problems with its plant, as well as water from the adjacent canal entering the factory discharge drain. Presumably these difficulties were overcome, for the factory continued in operation until July 1974. Two years later it was sold to Wilderness Gems for the large-scale marketing of gemstones and mineral specimens throughout New Zealand and overseas. The amount of space available in such buildings has facilitated new uses including craft shops, tearooms, an indoor cricket centre, woollen knitwear manufacture, a museum, and no doubt various others, all illustrating the flexibility of the disused dairy factory.

The year 1923 saw the spanning of the Taraheru River in Gisborne by a new Peel Street Bridge. It was designed by John Alexander Macdonald, Borough Engineer from 1919–23, and consists of several spans using reinforced concrete beams and deck slab. Originally the balustrades were solid but have been changed to an open metal pattern with vertical bars. In 1925 a similar design was used for the Kaiti Bridge (Gladstone Road) across the Turanganui River, about 400 m from the Peel Street Bridge. Both had a finely textured plaster finish to the balustrades but this has discoloured over the years. The design is straightforward and pleasing.

In marked contrast to the Kaiparoro Memorial Bridge, mentioned earlier in this chapter, is the beautiful Bridge of Remembrance built in 1923 at Christchurch to span the diminutive Avon at Cashel Street. The idea of a war memorial in bridge form was one of many put forward by citizens in response to the call of the Mayor, Dr T. Thacker, for a 'permanent peace memorial'. After much public discussion it was agreed to call for precise competitive proposals nationwide and the successful entry was that of Gummer and Prouse, a firm of architects in Auckland and Wellington, for a 'Bridge of Remembrance'. William Henry Gummer had already made his mark with several outstanding buildings and he had a particularly fine appreciation of reinforced concrete as evidenced at Tauroa (see pages 152–4)).

Using a single elegant 15-m segmental arch to span the stream at an angle of 30 degrees, he had as a vertical element a tall arch over the roadway at the east end. This had smaller flanking arches on the sides for pedestrians and above these were lions couchant carved in stone. The structure was faced in Tasmanian sandstone — no doubt many engineers, and architects too, would have shunned the idea of facing a utilitarian structure such as a bridge with expensive imported stone on the grounds of obscuring an honest expression of form and material. However, there is a long tradition of memorials being constructed of stone and the public would expect an impressive treatment, regardless of what structure had been chosen, to express the idea of peace.

The old bridge it replaced had been identified by the passage of thousands of soldiers marching from the nearby King Edward Barracks to the railway station en route to Lyttelton and the troopships for First World War reinforcements. The architects envisaged a bridge which people would use for leisurely strolling and enjoyment. (*To page 186*)

KAITI BRIDGE, GISBORNE.
Also known as Gladstone Road Bridge this was built over the Turanganui River in 1925. It followed the same design as the Peel Street Bridge built two years earlier, both the work of the Borough Engineer, J.A. Macdonald. The Kaiti Bridge retains the original solid balustrade but the one at Peel Street has been replaced by an open metal version.

BRIDGE OF REMEMBRANCE, CHRISTCHURCH.
This is a beautiful bridge and a fine war memorial. It was the successful result of a competition won by Gummer and Prouse and was opened in 1923 over the Avon at Cashel Street. The concrete is faced in Tasmanian sandstone as a concession to public expectations at that time. As a functional memorial it is superb and its conversion to pedestrian use has increased its aesthetic merit.

CARGO SHEDS, PRINCES WHARF, AUCKLAND.
This is a fine example of the flat slab method of construction with mushroom columns and no beams. There are six large two-storey connected sheds with concrete roofs designed by the Auckland Harbour Board's Design Engineer, N.L. Vickerman. Completion was in 1923–24. The photograph of the interior of these sheds shows the flat slab construction. Modern cargo-handling methods have made such sheds redundant.

Opposite: MAIN STAND, TRENTHAM RACECOURSE, UPPER HUTT.
S. and A. Luttrell were responsible for this stand and also for the stewards' and members' stands. Architect Sidney Luttrell was the designer and in this building he produced a superb structure. He exploited concrete to express the bold but integrated use of ramps and stairs at the rear.

Freezing works, Hicks Bay, East Coast.
Now dilapidated, this building was erected in 1921 for local farmers in a remote district, with misplaced optimism — the works closed in 1926. The photograph shows the slender columns supporting arched beams, giving a suprisingly graceful effect for such a utilitarian building.

Opposite: Church of St Mary of the Angels, Wellington.
This impressive church in the heart of the city is one of the finest in the land. Designed by F.de J. Clere in 1919 it was eventually completed in 1922. The predominance of reinforced concrete is seen in the slender soaring arch ribs supporting the roof. The finely proportioned twin towers of the west front and the large rose window give added character to this national treasure.

STEWART CREEK BRIDGE, ISLAND BLOCK, CENTRAL OTAGO.
This little bridge was erected in 1922 as a memorial from the Stewart family to their parents and was gifted to the Tuapeka County Council. It is a rare, if not unique, memorial in New Zealand.

MEMORIAL BRIDGE, KAIPARORO, WAIRARAPA.
Built in 1922 this bridge is an Anzac memorial. It was designed by A. Falkner for the Mauriceville County Council. The small, shallow arch is dominated by the solid balustrades so that the usual grace of the arch is greatly diminished.

MEMORIAL GATES, TUAPEKA WEST.
This war memorial was erected in the early 1920s as an entrance to a former school.
The effect now is one of loneliness and mystery.

SURGE TANK, TARIKI POWER STATION, TARANAKI. Located on the Manganui River this station came into operation in 1924. The octagonal surge tank is a picturesque reminder of this small scheme, which was soon superseded by that at Motukawa Powerhouse in 1927.

This idea has come almost full circle with the road traffic now being diverted to a new bridge and the Bridge of Remembrance having been widened and restricted to pedestrian use in a pleasing mall-like treatment.

There are many other war memorials throughout New Zealand commemorating the fallen in the New Zealand Wars, Boer War and both World Wars. Most are of stone or with stone facings, but sometimes concrete was used and expressed honestly. Generally this material was considered to be too utilitarian and lacking a suitable texture for dressing and working up or polishing.

One interesting example can be seen at Tuapeka West in Otago as memorial entrance gates at a former school. Passing that way one foggy morning, my wife and I were surprised to see this monumental gate to nowhere giving access to a paddock sloping downhill. Long grass and disuse prevented us from opening it. The form of this memorial consists of short concrete walls about a metre high flanked by square concrete posts with inscribed memorial stone plaques. The walls extend at 45 degrees to join a most unusual horseshoe concrete arch over the single-leaf wrought iron gate. The apex of this arch has another inscription and is surmounted by a dove, symbolising peace; unfortunately it has been damaged. Enquiries through the Tuapeka County Council did not establish a date, but like most First World War memorials it was most probably built in the early 1920s.

During this decade hydro-electric development made considerable progress throughout the country. There were three state enterprises and a number of small low-head schemes initiated by local authorities and power boards. One of these was designed for the Taranaki Electric Power Board by H.W. Climie and Sons, consulting engineers, and approved in 1923. Henry Westcott Climie had designed the dam and headworks for the Stratford Borough hydro-electric scheme which began generation in 1901 — the first public scheme in the North Island using hydro-electric power. It no longer exists except as a very small concrete ruin.

By August 1924 the Tariki hydro-electric station was commissioned and power generated for local consumers. The scheme consisted of headworks on the Manganui River near Tariki with water conveyed by a small concrete-lined race to a generating plant as a first stage. The concrete surge chamber, octagonal in shape, can still be seen on the riverbank. Work continued on a series of races and tunnels to form Lake Ratapiko and from there the water was brought by tunnel to a surge chamber 20 m high and 9 m in diameter. Steel penstocks took the water to the nearby Motukawa powerhouse with the tailrace to the Waitara River. This scheme was opened in 1927 and is still in use, although some renewal work has been undertaken following major floods. A series of lightning strikes necessitated re-winding the generators in 1969.

We noted in chapter one that the beamless floor slab was developed by C.A.P. Turner in the United States in 1908 and refined by Robert Maillart as the better-known mushroom column or flat slab construction. One of the best known examples of the Turner version was in the Van Nelle factory built in Rotterdam in 1927–28. However, a large-scale use of this method was carried out by the Auckland Harbour Board with the Princes Wharf cargo sheds, which predated the Dutch building by four years.

In a brief dated 29 April 1919 the board's Design Engineer, N.L. Vickerman, was instructed to design the Hobson (later renamed Princes) Wharf in reinforced concrete with a central roadway of 18.28 m flanked by double-storey sheds 24.38 m wide. The space between the six sheds is also 24.38 m. The wharf deck is supported on piles of 457 mm by 457 mm up to 18.28 m long and 508 mm by 508 mm where the length is greater. With a wharf measuring 345.3 m in length and 43.5 m wide there is ample berthage on both sides for large vessels.

Each of the six cargo sheds is 97.5 m long by 24.38 m wide. The ground floor has a height of 3.8 m with 609-mm-diameter columns, whereas the first-floor columns are 457 mm in diameter and the floor height is 3.38 m. These circular columns have conical heads and square drop panels (slabs), but the perimeter walls include rectangular columns having splayed column heads with asymmetrical drop panels supporting the cantilevered upper floor slabs. These provide canopies along the exterior. The six sheds have concrete roofs and were built in 1923–24.

The first pile for the wharf was driven in October 1921 and when the project was completed in 1924 a grand official opening was held in the presence of the Governor-General, Lord Jellicoe. Huge crowds pressed close to the sheds to admire HMS *Hood*, Britain's 41,656 tonne battlecruiser which was berthed alongside.

The construction of the wharf and sheds was supervised by N.L. Vickerman on behalf of the Chief Engineer, W.H. Hamer, using day labour. There can be no doubt that great care was taken throughout to ensure high standards of

DEVONPORT FLATS, NEW PLYMOUTH.
Located on the corner of St Aubyn and Dawson Streets this was one of the largest blocks of flats in New Zealand when completed in 1924. They were designed by F. Messenger, a leading Taranaki architect. The elevations have unity and liveliness.

workmanship, for when I last visited in June 1987 these structures were in remarkably good condition. Some redevelopment was planned by the harbour board but the wharf deck will be retained and ideas will be examined for new uses of the sheds, which may be kept whole or in part. This remarkable group is a splendid example of a well-designed concrete industrial project.

What must have been one of the largest blocks of flats in New Zealand in the early 1920s was erected in New Plymouth. The Devonport Flats were built for A.B. Waldie on the corner of St Aubyn and Dawson Streets. The architects were Messenger, Griffiths and Taylor of whom the senior partner, Frank Messenger, was a leading Taranaki architect for many years. Designed in two stages, the plans for the flats are dated May 1922 and December 1922, with completion of the building taking place in 1924. Of four storeys, the Devonport Flats are of reinforced concrete construction with 127-mm-thick floor and roof slabs supported on beams. The elevation to St Aubyn Street has the two lower floors integrated with large round-headed windows having spandrels at first-floor level. The Dawson Street end has a more restrained appearance with bays above ground floor level, but there is a unity in the design which is both lively and disciplined.

At various times the Wellington City Council has been responsible for some good civic architecture and engineering in concrete. One example is the Basin Reserve Pavilion built in 1924. The foundation stone records the name of the City Engineer, A.J. Paterson, but there is no mention of the architect. The rear elevation in Sussex Street is certainly a most pleasing one, showing a classical treatment with large round-headed windows. The seating in the stand is supported on reinforced concrete and the roof has steel trusses. As was customary for most public buildings in this decade, raw concrete was plastered in a cement render to make a more acceptable visual finish.

Also in 1924 a concrete arch bridge was opened in Stratford. It spans the Patea River on Broadway, the main street, and was designed by the Public Works Department under F.W. Furkert as Engineer-in-Chief. It has a 19.5-m arch, which must

BASIN RESERVE PAVILION, WELLINGTON.
The older of two pavilions, it was built in 1924 under the City Engineer, A.J. Paterson. It is a pleasant building, particularly from the rear bordering the one-way street. The neo-classical treatment has large round-headed windows. The concrete surfaces are cement rendered throughout.

make it one of the widest concrete bridges of its era in the country.

In 1924 F.C. Daniell, whose work in reinforced concrete was noted earlier, made substantial additions to a single-storey block in Victoria Street, Hamilton. This was to be his largest commercial building in that town and consisted of two more storeys to give a handsome facade to the main shopping thoroughfare. The original portion was designed to have the additional floors. The central part is recessed, with a cantilevered balcony having a metal balustrade at the first-floor level. The slightly projecting portions have central bay windows for the upper floors. The design is asymmetrical with large windows in the centre and north end. This fine building, known as Wesley Chambers, had until recently a facade of plastered brick but the rendering has been removed to expose the brick.

The 1920s does not seem to have produced any marked increase in the number of houses built of reinforced concrete. As early as July 1914 a patent was granted to William Percy Glue, a painter of Timaru, and his carpenter brother, Walter Frederick, for 'an improved collapsible mould for the erection of concrete walls'. It used steel-lined boards as a quick and inexpensive method of constructing either a hollow or solid wall in concrete. The moulds for the outside could be adapted to swing outwards and the inside moulds adapted to swing inwards. From 1916 the brothers built houses with hollow concrete walls in Timaru, Dunedin and Christchurch. In 1926 W.P. Glue used a hollow concrete wall system comprising two 75-mm exterior walls having a cavity for reinforcing. The internal walls were 100-mm solid concrete. The Glue Construction Coy Ltd of Dunedin built houses of concrete throughout the 1920s.

The first state hydro-electric power station in the North Island was the Mangahao, opened in November 1924. The powerhouse, built of reinforced concrete, is only a few kilometres from Shannon in the Manawatu district. Like its predecessor at Lake Coleridge, this has a straightforward approach with its elevations dominated by the large metal-framed windows for natural lighting. In spite of these being square-headed, the powerhouse has something of the Gothic structural response in the relationship of solid to void. The station began with two generators, and three more were added by mid-1925 to give 20,000 kW of power. The water is stored in three reservoirs reached by a narrow, tortuous and steep road.

The dams are concrete and a tunnel links Nos 2 and 3 reservoirs. Another tunnel takes the water from No. 1 dam to the surge chamber and thence by steel penstocks to the turbines in the powerhouse. A small power station by post-Second World War standards, it was a pioneer project and a valuable link in the national power supply.

Two years after the Mangahao scheme, a small station was opened at Piriaka on the Whanganui River for the Taumarunui Borough Council. Here the surge chamber is a circular concrete tank alongside the powerhouse, which is a simple gable-roofed building of concrete.

A town bridge designed in 1926 by the Auckland firm of Jones and Adams is still in use in Te Aroha. The Waihou River Bridge has ten spans of 12.19 m and has the axes of the roadway at 60 degrees to the piers, making it a skew bridge. These piers are concrete with a core of cyclopean concrete, an infill concrete having aggregate larger than 150 mm. The piers have rounded columns at the ends and rest on piles, with the maximum length being 12.3 m. Slightly haunched beams carry the concrete deck with pedestrian ways on both sides. One of the hallmarks of many concrete bridges designed by Jones and Adams is the floriated pattern in the balustrades. Here each parapet span has eleven squares containing modified quatrefoils. The original design showed graceful lampstands with special lights but these are not in place now.

The history of bridge building in New Zealand is rich in human interest. One bridge with a most unusual origin is that at Kawarau Falls near the Kawarau River outlet to Lake Wakatipu. In 1922 a longstanding intention to construct a bridge began to be realised when concrete pier bases were built during a particularly low winter level of the river. However, in January 1923 there was a proposal to put a dam in the river for dewatering so as to allow supposedly rich recovery of the gold from the rocky nooks and crevices of its bed. This idea caused a delay by the Public Works Department in continuing the bridge construction and a year later the Kawarau Gold Mining Company was granted a licence for a dam. There was a proviso that there should be a bridge along its crest. Many objections were lodged to the licence being issued, but subsequently there was a rush of applications for prospecting licences.

The company's engineer, E.J. Iles, estimated the cost at £30,000. His first task was to remove several small islands of rock by explosives and then to demolish the concrete pier foundations. Work began in late December 1924 and was completed two years later. In 1925 the consulting engineers Vickerman and Lancaster of Wellington replaced Iles, who had left the company. The construction of the dam was not only protracted but it cost the staggering sum of £100,000 — a very large amount for that time. The Kawarau Falls Bridge, as it is known today, has a length of 143.6 m, is 3.6 m wide and 10.6 m above normal water level. The steel girder spans rest on eleven reinforced concrete piers, their leading edges being raked on the upstream side. Between them the sluice gates were fitted.

WAIHOU RIVER BRIDGE, TE AROHA.
Jones and Adams were the engineering consultants for the borough council. The bridge has ten spans of 12.2 m and is a skew bridge with the roadway axis at 60 degrees to the piers. As with some other bridges, the designers have used a concrete balustrade pierced in a modified quatrefoil pattern. The design was prepared in 1926.

STEPS, UPPER DOWLING STREET, DUNEDIN.
This fortress-like structure in the heart of the city was erected in 1926 when the street was widened. It has a ramp and is also a large retaining wall.

When the great day came for closing the gates thousands of spectators arrived to see what the dry bed below the dam would reveal. There was great disappointment however, especially by the Kawarau Gold Mining Company, when it was realised that the water found its level in the Shotover River mouth about 6 km upstream to form a lake. Below the dam a range of rock bars across the riverbed acted as locks to retain the water. It was also discovered that earlier sluicing operations had left large quantities of debris in the river, which tended to retain the water at a higher level. Several other factors added to the general dismay. The 1926 closure produced only a very small amount of gold, but in the winter of 1927 a further closure of the gates allowed rather more gold recovery, though not in the quantity envisaged. While the dam failed to achieve its intended aim it certainly provided a satisfactory bridge which is still in use on a main tourist highway.

The Napier earthquake of 3 February 1931 caused most of the old brick buildings in the district to collapse in ruins. Even some of the reinforced concrete framed structures with brick infill panels showed considerable damage. One of these was the Public Trust Office building in Napier situated on the corner of Tennyson and Dalton Streets. The Greek Doric design was the successful entry of Eric Phillips, a local architect, in a competition which saw its construction completed in 1926. The structure is of reinforced concrete frame with the original walls being brick panels. Extensive damage to this brickwork necessitated its removal after the 1931 quake and replacement in reinforced concrete. The original working drawings show the reinforcement for the columns as train rails. This is a fine building which is unashamedly classical — a house style of many of the earlier Public Trust Offices. Because its structural frame remained intact after the earthquake, while so much around it lay in ruins, this building became something of a symbol of faith in the virtues of reinforced concrete design.

Retaining walls are a feature of Wellington, especially in the hill suburbs where streets sometimes have high sidling cuts to provide suitable alignments. Dunedin also has some interesting retaining walls with several in the central city area. As far back as 1886 Upper Dowling Street was being formed by a cutting above Princes Street. This streetwork was noteworthy for a tragic blasting accident when two people were killed and several injured. In 1926 this portion was widened and an impressive ramp built with steps and a massive retaining wall. The steps are enclosed in a citadel-like structure that takes on the appearance of the local basalt in its rather gloomy coloration.

A splendid example of the work of William Henry Gummer is the large house known as Arden in Havelock North. It was built in 1926–27 for Maurice Chambers, son of Mason Chambers who had commissioned Gummer to design his Tauroa house, already described on page 152.

Arden is sited to give a fine view from the elevated portion of the town. It has a reinforced concrete frame and floor with cavity brick infilled panels as used by Gummer in Tauroa and in the Craggy Range house, now known as Belmount (page 162). A large building of two storeys with an area of approximately 2,400 sq. m, Arden gives an impression of Mediterranean domestic architecture. This is emphasised by the white plaster finish. The main garden elevation has steel round-headed French doors at ground level and rectangular steel windows in the upper storey with a balcony. The courtyard at the rear has a service wing of two storeys with its own exterior staircase in concrete. Near the intersection of the two wings is a delightful oriel window. A recently added porte cochère at the main entrance has been skilfully executed to blend with the original design. Now a guest house, Arden displays the architectural skills of W.H. Gummer in all of its features.

ARDEN, HAVELOCK NORTH.
Using the same construction as for Tauroa and Craggy Range, W.H. Gummer designed this fine house to give a suggestion of the Mediterranean. The rear is also attractively designed and the house has a fine garden setting with a view. It was built in 1926–27.

Concrete was used as a tunnel lining in the 1890s on the Kohatu No. 1 railway tunnel on the Nelson–Glenhope line. Although the Karori road tunnel uses brick, the nearby Northland Tunnel constructed in 1927 has a concrete lining, as have all subsequent Wellington tunnels. At its northern approach it has a concrete portal with splayed wing walls sloping from the top to a little over a metre above the ground. The tunnel is parabolic in profile.

In 1926 a concrete lighthouse was built at Cape Foulwind to replace the old wooden tower of 1877. A small structure built on the granite rock of the cape, it comprises a plain, windowless tower surmounted by a beacon, with a catwalk that barely protrudes beyond the wall. It is now automatic, as are all New Zealand lighthouses.

In the nineteenth century there were various do-it-yourself efforts using mass concrete. By the mid-1920s this wish to be self-sufficient in design and construction of concrete structures was still evident on occasions. Two instances, both small churches in isolated communities, are worthy of mention. In the area north of the Hokianga Harbour, the Mill Hill fathers of St Joseph's Missionary Society of London had been active in the latter part of the nineteenth century and again in the 1920s. Their early churches, such as St Gabriel's at Pawarenga on Whangape Harbour, were built of timber. When Fr Charles Kreijmborg arrived in this district he set about building a church at Rotokakahi. He had come to New Zealand in 1893 and began his ministry at Matata in the Bay of Plenty where he completed a timber church two years later. In 1896 he went to Tokaanu-Waihi on the southern shore of Lake Taupo and added a sacristy to the fine timber church.

When Fr Kreijmborg moved to Rotokakahi he decided to build in reinforced concrete and on 6 February 1927 the Church of Our Lady of Lourdes was opened for services. It was certainly a bold achievement by one who was architect, clerk of works, foreman, masterbuilder and a co-worker with several Maori helpers. Before his arrival in New Zealand Fr Kreijmborg had been thoroughly trained as a builder and carpenter and it says much for his energy and determination that he used reinforced concrete without skilled tradesmen. What is more remarkable is his degree of understanding of how to design and place the steel reinforcing — one suspects it was very elementary, although the fact that the two buildings are still in use today testifies to his ability. This church has timber roof construction with an attractive ceiling. The exterior presents a pleasing appearance with its recessed full-width porch behind triple arches. The front elevation is flanked by small twin towers above roof level. There is now a convent and school alongside the church.

At the small settlement of Waihou, not far from Panguru on the Hokianga Harbour, Fr Kreijmborg built another reinforced concrete church dedicated to SS Cletus and Remigius. This was opened for worship on Easter Day in 1928 by Bishop Cleary. Here the ceiling is less elegant than at Rotokakahi, the roof being supported by concrete columns cast in metal forms but with wooden corbels at the top. There is a small gallery at the rear. The front elevation also has a recessed porch but it is not full width. Two square columns with non-classical capitals form a three-bay opening. Above this the wall terminates in a broadly stepped fashion with surmounting pediment and arched support for a cross. Fr Wanders, whose grave is near the entrance, was heard to remark that this church was more successful structurally than Fr Kreijmborg's first effort.

The city of Wanganui has a notable landmark on Bastia Hill in the form of a tall water tower completed in September 1927 by the City Engineer's Department. It was designed by E.A. Gumbley under the City Engineer, N. Crofton Staveley, and is an open structure — that is, there is no solidly enclosed tower in the shaft but a series of raking columns to support the water tank at the top. These columns are stiffened horizontally by three decks with pierced balustrades. There is further lateral bracing in the form of graceful arches with vertical stiffening struts. A central core has a series of spiral concrete stairs for access. The water tank of 544,800 litres capacity repeats the theme of flowing curves by having a shallow arcade and a domed roof. Without doubt this 50.3-m-high tower is an intriguing design expressing a happy marriage of grace with utility in fairfaced concrete. It is a structure commanding admiration.

By the late 1920s there were some significant 'high rise' city buildings in Auckland and Wellington. High rise was limited at that stage to 102 feet (31.1 m) in the capital for seismic reasons. One such is the T & G Building on the corner of Lambton Quay and Grey Street, built in 1928. The Temperance and General Mutual Life Society engaged their Australian architects A. & K. Henderson for this eight-storey building which displays the transitional approach to much commercial architecture at this time. It has a splendid corner treatment and the structure is given expression in the form of alternating banks of recessed windows having bronze spandrels and flush strips of wall between, with the windows as pierced holes. There is a strong capping to the roof line with the projecting cornice decorated with modillions. The entablature below has minuscule paired Doric columns flanking round-headed windows. These look onto balconies with open balustrades. In spite of the classical elements there is a modern approach to the facade treatment and a feeling of concrete rather than brick or masonry. The T & G Building, with its sculptural quality, makes very good streetscape. (*To page 202*)

OUR LADY OF LOURDES CHURCH, ROTOKAKAHI, HOKIANGA.
This church is testimony to the abilities of the Mill Hill priest, Fr Charles Kreijmborg. He used reinforced concrete, doing all the design and construction himself with a few Maori helpers. The building was opened for services in February 1927.

CHURCH OF ST CLETUS AND ST REMIGIUS, WAIHOU, HOKIANGA.
Fr Kreijmborg built another church, in 1928, also in reinforced concrete. This bears little resemblance to traditional churches in New Zealand.

PUBLIC TRUST OFFICE, NAPIER.
E. Phillips, a local architect, used a reinforced concrete frame with the original walls of brick panels in this competition-winning design. Completed in 1926, the panels had to be replaced in concrete after the Hawke's Bay earthquake of 1931. This fine structure complements the later Art Deco buildings of the early 1930s.

Kawarau Falls Bridge, Central Otago.
This structure has a convoluted history. The resultant bridge incorporates a dam with sluice gates which were intended for dewatering the Kawarau River below it for the recovery of alluvial gold. The cost was prodigious. There are eleven concrete piers for the steel superstructure, with sluice gates fitted between. The work, which began in December 1924, was completed in 1926 and the engineers for most of it were Vickerman and Lancaster of Wellington.

CAPE FOULWIND LIGHTHOUSE.
Although a small structure, this is of note as being the first poured *in situ* concrete lighthouse in New Zealand, dating from 1926. It is seen at close quarters by those using the Cape Foulwind Walkway.

Water tower, Bastia Hill, Wanganui.
This tower breaks the tradition of such structures being enclosed. Here there are raking columns, open decks, varied use of arches and a crowning dome to suggest flowing movement and grace. The work of the City Engineer, N.C. Stavely, and E.A. Gumbley, this project of 1927 is a landmark in the city.

Opposite: AMP Building, Customhouse Quay, Wellington.
F. de J. Clere is regarded as an ecclesiastical architect but here he shows his mastery of a large head office block. Designed in 1925 it was completed in 1928. Restrained and dignified, this is a fine piece of commercial architecture.

T. & G. Building, Lambton Quay, Wellington.
Designed by the society's Australian architects, A. & K. Henderson, this 1928 building emphasises the verticality of the columns with the playing down of the dark bronze spandrels. The facade expresses a modern feeling in spite of some lingering classical detail.

In the same year as the T & G Building was completed the Australian Mutual Provident Society (AMP) opened their rather larger national head office building in Customhouse Quay in Wellington. Designed in 1925 by F.de J. Clere it is a splendid example of the newer technology wearing the dignity of a scholastically sound classical treatment in decorative elements. A large office building for its time, it has six storeys of office space and a semi-basement to give a raised ground floor. It is on a corner site, with the external appearance being reminiscent of the Renaissance palazzi in Italy — a style that influenced much public and commercial architecture of the 1920s in the United States.

The AMP building has one of the grandest entrances of any in New Zealand for its period. The steps lead to a vestibule deeply coffered in its barrel-vaulted ceiling and having richly veined marble walls. As an expression of reinforced concrete this building displays the traditional approach of finance houses and insurance companies — a clinging to classical and formal qualities in architecture and for good measure clothed in Hawkesbury stone from Australia on a granite base. It was 41 years since Clere first used concrete at Overton, for he had been an early exponent of its structural virtues. No doubt the client was dictating the requirement for the proven dignity and assurance of a classical building. Clere, always an architect of great refinement, created a superb building of classical integrity and balance.

An interesting wharf structure is that at Tolaga Bay designed by Cyrus John Richard Williams when he was engineer to the Tolaga Bay Harbour Board. Construction was completed in 1929 although it began three years earlier. Parochial pride allowed the wharf to become an overly ambitious scheme for such a small settlement and the cost was never justified, with Gisborne only 54 km distant. Nevertheless it is a striking monument with its tremendous length paralleling the nearby cliffs on the south side of the bay. It hasn't been used for some time other than for fishing and the steel is showing signs of severe corrosion where the concrete cover has spalled off. Visually it is a satisfying design with its raking piles and balustrades of concrete posts with rails of the same material.

Several hydro-electric generating schemes have already been mentioned. The largest in the North Island in the first three decades was the Arapuni project on the Waikato River about 18 km from Putaruru. The concrete 'arch' dam spanning a deep gorge with sheer walls of rhyolite breccia is 64 m high. It was the largest dam to be built for hydro-electric purposes in New Zealand at that time. The contract was let to the British firm of Sir W.G. Armstrong, Whitworth and Company after world-wide tenders had been called. Work began in September 1924.

The principal concrete structures are the dam, spillway, penstock intakes and powerhouse. Although curved, the dam is a horizontal gravity type relying essentially on its weight and strength, with the curved form as an additional safeguard. The scheme came into partial operation in June 1929 and the first stage was regarded as complete in 1930. The powerhouse has steel framing encased in concrete in the generator room, with walls, floor and roof of concrete.

The Arapuni project was controversial from the start. A great deal of apprehension was felt by some people as to the safety of the dam and some local downstream farmers even wished to sell their properties and leave the district. The contractor was criticised severely for the lengthy delays, and more particularly for incompetence in handling the work. In June 1930 construction had to stop when it was discovered that major cracks in the cliff alongside the powerhouse were allowing water into the building. After close geological investigation it was found that a block of country near the penstock tunnels and between the spillway and powerhouse had broken away on 7 June.

Advice from world authorities was sought by the government, and Professor P.G. Hornell and P.N. Werner from Sweden were appointed to report on the exact cause and suitable remedial measures. Their report was adopted by the government and authority was given in December of that year for the Public Works Department to proceed with the necessary work. The country rock was strengthened with steel rods grouted in under pressure. This use of cement grouting was to become widely used in the ignimbrite rock abutting dams upstream built in the post-Second World War years — a technique perfected by the Ministry of Works when it was formed from the Public Works Department in 1945.

Other major construction work was carried out by the PWD at Arapuni before the station became operative again, and most of this involved considerable use of reinforced concrete. The station recommenced operations in May 1932, with four generators installed. The power-house was extended in 1938 to give a total capacity of 157,000 kW from eight generators. The Arapuni scheme was certainly a major civil engineering undertaking for the time. It involved immense technical difficulties and reinforced concrete played an essential role. The government learned that the PWD had the ability to design and construct major power schemes and this became the general policy from that time. It is interesting to note that in many projects where overseas contractors have been involved the PWD (and later the MOW) has either been asked to complete the work or else act as nursemaid. The Arapuni project called for a construction village with houses and facilities for the operating and maintenance staff. In 1926 a concrete water tower was erected for their needs — a cylindrical tank supported on horizontally braced reinforced concrete columns. It makes an interesting comparison with the form used for the water tower built in 1883 at Addington Railway Workshops.

While construction was going on at Arapuni, the state's second South Island hydro-electric dam was begun in 1928 on the Waitaki River about 6 km from Kurow. The concrete dam was a curved gravity type, 36.57 m high, built on faulted

Town Hall, Hunterville.
This small country-town public building is understandably modest. There is a semblance of Art Deco in the elevation above the verandah roof.

Retaining wall, Plunket Street, Wellington.
This sinuous wall in Kelburn was designed in the city engineer's office and constructed in 1929. There is little attempt at any form of embellishment. A high level footpath, a feature of Wellington hill suburbs, can be seen behind the concrete posts and pipe rails.

greywacke argillite rock. It was the first in the country to be built across a major river where its diversion was not practicable during construction. It was also built by a large workforce mostly using picks, shovels and wheelbarrows — some 1,200 men were employed at the peak period. Disposition of the material was effected by using light gauge railway lines with spoil trucks running on them. This project has links with the origins of New Zealand's social security system through the development of the Waitaki Hydro Medical Association and the work of Dr D.G. McMillan, who later entered Parliament. The Waitaki project came into operation in 1935 with the completion of the reinforced concrete powerhouse generating initially 75,000 kW and later 105,000 kW of electricity. The dam is unusual in New Zealand in not having a spillway but allowing the water to flow freely over the crest when necessary. The importance of salmon fishing on the Waitaki later necessitated a fish ladder as part of the dam structure.

In the North Island another small hydro-electric power station was built in the late 1920s using water from Lake Waikaremoana. The Tuai powerhouse was commissioned in 1929, being a small reinforced concrete building. Because water was drawn from the lake, a dam was not required.

Retaining walls in cities have been mentioned earlier. One of the older extant walls in Wellington is in Plunket Street, in the residential suburb of Kelburn. This gently sinuous wall of the reinforced concrete cantilever type was built by contract in

FORMER PUBLIC TOILET AND TRAM SHELTER, COURTENAY PLACE, WELLINGTON.
This amenity was saved from demolition by public demand. Built in 1927 it was popularly called 'The Taj Mahal' because of the apsidal domes at each end. In recent years it has been a restaurant and café, with several changes of owner and colour scheme.

1929, having been designed in the office of the City Engineer, George A. Hart. The main reinforcement is of 9.5 mm rods at 460-mm centres. The wall has a plain, off-the-form surface without patterns or dividing strips of any kind and it has a slight batter on the face. The top, with its projecting cap, is supported on concrete corbels, above which is a handrail consisting of concrete posts carrying metal pipe rails. There are steps at the north end parallel to the wall.

An example of a small country town public building is the Town Hall in Hunterville, built in 1929. It is a free-standing two-storey building with an unpretentious verandah canopy. The front elevation has a row of round-headed windows with

Opposite: TOLAGA BAY WHARF.
Construction began in 1926 for this long wharf but it was not completed until three years later. It was designed by C.J.R. Williams, Engineer to the Tolaga Bay Harbour Board. This was another over-ambitious local scheme. Severe corrosion of the steel is appearing in the piles.

150 CLYDE ROAD, CHRISTCHURCH.
The brothers W.P and W.F. Glue built a number of reinforced concrete houses using a patented system of collapsible moulds. The houses were erected in Dunedin, Timaru and Christchurch in the 1920s and 1930s. This one in Clyde Road was built in 1929 and is still in very good order.

pilasters between them giving an Art Deco effect. The building is of column-and-beam framing with concrete infill panels and a trussed roof.

Over the years some structures attain strong local identity in the public estimation and a few receive sobriquets befitting their particular character. One of these is the former public toilet block in Kent and Cambridge Terraces, Courtenay Place, Wellington. As a result of a student prank it was nicknamed the 'Taj Mahal'. This stuck and was often shortened to 'The Taj'. Although the building has rather flat cupolas (domes) and rounded apsidal-like ends there is no resemblance to the renowned and magnificent Agra monument. The nickname has remained even with subsequent changes of use (and name) as art gallery and patisserie, café restaurant and gallery, and a café.

'The Taj' was built in 1927 in reinforced concrete by the Wellington City Council but there is no record of the architect. It was designed as a tram shelter and public toilet with provision for both sexes, and had a very high standard of interior finishes. The patterned tiling was superb with pleasant summery green hues. There is no doubt that the round apsidal ends with their domes provided interest in their attractive forms. When buses replaced trams and new amenities were provided, the building became redundant and was condemned for demolition. However, strong public concern and pleas for its retention has ensured its survival.

Dam, Arapuni.
Sited on the Waikato River, this curved horizontal gravity dam is part of a large hydro-electric project plagued by problems during the prolonged construction period. It is 64 m high in a gorge with vertical walls of rhyolite breccia. The dam was completed in 1930.

Opposite: Dam and powerhouse, Waitaki.
This was the first hydro-electric scheme on the Waitaki River. It began in 1928 and is sited near Kurow. The dam does not have a spillway but water can flow over the crest when necessary. It is a curved gravity structure 36.57 m high. The project was completed in 1935.

Chapter Ten
1930–39

In the fourth decade of the twentieth century reinforced concrete construction was in widespread use and certainly much more popular than structural steel. Concrete was a material the small contractor had learned to handle and in which there was growing confidence. Cement was produced locally, reinforcing steel was readily obtainable from Australia, and suitable aggregates both as sand and gravel were fairly well distributed. It would seem that the adaptable Kiwi had already become familiar enough with concrete for do-it-yourselfers to carry out many small works about the property such as paths, drives, kerbs, foundations for garden sheds and garages, and garden walls. Nevertheless this interest was modest when compared to the post-Second World War boom in such projects. By the 1930s there was so much concrete construction in all manner of forms and usage that only a few representative examples can be mentioned to illustrate trends.

The Auckland Railway Station is a reinforced concrete building faced in brick, but a more obvious expression of the structural qualities of the material may be seen in the train platforms at the rear. The architects for the station complex were Gummer and Ford of Auckland and the design of the canopies over the platforms illustrates their aesthetic control of all aspects of design and construction. These canopies date from 1930 and have central columns with a longitudinal beam supporting cantilevered beams and roof slab to exploit the station roof principle.

As mentioned in the previous chapter, an early firm of consulting engineers in the twentieth century who produced some interesting concrete design was Jones and Adams of Auckland. Stanley Jones had a brother, Gerald, who was an architect enjoying a reputation for his sensitive house designs. There can be no doubt that both were gifted in their respective disciplines. In 1930 the opening took place of the Jones and Adams-designed Ngamuwahine Bridge over the river of that name on State Highway 29 in the Bay of Plenty. Of three spans, it has a balustrade of an open floriated pattern to give a feeling of lightness and grace to an otherwise rather utilitarian structure. The bridge has a gentle rise to the centre, with the beams being haunched over the piers and abutments. It was built for the Tauranga County Council with a £2 for £1 subsidy from the Main Highways Board, the body established in 1922 as the antecedent of the National Roads Board and what is now Transit New Zealand. The floriated design of the balustrade was also used by Jones and Adams in other bridges, as at Te Aroha, Ryburn's Bridge near Clevedon, and the Puhinui Stream Bridge at Wiri.

The need for earthquake resistant buildings was of paramount importance to the Council of Governors of Nelson College. After the destruction of a large part of the main building in the 1929 Murchison earthquake it was decided that the first priority for rebuilding would be three new houses for the boarders. The first two were named after Baron Rutherford and the Hon. J.W. Barnicoat. The foundations for these houses were laid in November 1930 and by September in the following year

PLATFORM CANOPY, AUCKLAND RAILWAY STATION.
The architects for the station were Gummer and Ford of Auckland. This use of reinforced concrete to create a cantilevered roof was to give a simple and maintenance-free form of construction. The station opened in 1930.

they were in use with provision for 60 boarders in each. The architect was W. Gray Young of Wellington.

Barnicoat House has the more commanding aspect, being on the lower slopes of Flaxmoor, and a short distance below it Rutherford House nestles sedately into the well-planted grounds. The two buildings, identical in plan and form, are constructed of reinforced concrete beam-and-bearing wall with a Spanish Mission influence in the architectural design. The form consists of a three-storey lower wing with a two-storey link and upper wing. As a boarder in Barnicoat House for four years, I recall the feeling of solidity produced by these H-shaped blocks with their robust concrete construction.

RUTHERFORD AND BARNICOAT HOUSES, NELSON COLLEGE.
When the 1929 Murchison earthquake destroyed part of the main school building, two identical boarding houses were built. They were completed in 1931 to the design of W. Gray Young of Wellington. The photograph shows Rutherford House with its Spanish Mission-influenced design, and open air dormitories on the upper floors.
(*Photo: A.R. Kingsford*)

BREW TOWER, TUI BREWERY, MANGATAINOKA.
Gummer and Ford of Auckland were the architects for this fine building begun in 1931. The reinforced concrete construction has partial brick facing and cement plaster. The origins of the brewery go back to 1889. Today the brew tower is a museum piece complete with its copper kettles and mash tuns.

Most of the early breweries were built in timber, stone or brick. The present DB Central Brewery at Mangatainoka in the Wairarapa had its origins in a small timber building in 1889, with brick structures following soon afterwards. In 1931 construction began on the fine brew tower designed by the Auckland architects Gummer and Ford. Of reinforced concrete with brick and plaster finishes, it has five storeys in a modified classical idiom. The hipped roof of corrugated galvanised iron has a lantern along the ridge. Verticality is given considerable emphasis by the use of brickwork on the corners and in narrower strips between the windows. 'Tui Brewery', the name of the company after 1923, appears along the frieze. Dominion Breweries made extensive additions and put in new plant in the 1970s but have left the brew tower as a museum piece, retaining the copper kettles and mash tuns. In its rural setting alongside the Mangatainoka River this structure is a fine landmark in the district.

The catastrophic earthquake of 3 February 1931 meant that Napier had very few buildings standing intact in the central

CENTRAL HOTEL, NAPIER.
Sited on the corner of Emerson and Tennyson Streets this splendid example of Art Deco was designed by local architect E.A. Williams in 1931. Construction soon followed and the two-storey hotel made a strong visual impact on the newly emerging streetscape of the earthquake-ravaged city.

CENTRAL FIRE STATION, NAPIER.
This was built in Tennyson Street in 1921, mainly of brick and steel with a minimum of reinforced concrete work. Much damaged in the earthquake, it was largely rebuilt in reinforced concrete, and with strengthening of the end wings. Now known as the Desco Centre and refurbished in the 1960s as professional offices, it has a pleasing frontage on Clive Square.

business area in the early 1930s. The Napier Town Planning Committee and the voluntary association of local architects pooled ideas and resources. It would have been easy to put up cheap and temporary buildings with a moratorium on permanent structures because of urgency and the shortage of money in the Great Depression. However, a spirit of optimism, the stimulus of new architectural ideas such as those of the Art Deco Movement, and the desire to make a superior town saw the proud achievement of 'the newest city in the world' well before the decade had ended.

One of the most important results of the 1931 earthquake, and also the 1932 quake which had such disastrous effects in all other East Coast and Hawke's Bay towns, was the preparation of the Model Building By-Laws as a national standard for all local authorities to adopt. Thus reinforced concrete, and to a lesser extent structural steel, became obligatory construction systems for public and commercial buildings in the central business areas.

A building signifying Napier's renascence is the Central Hotel, occupying a corner site on Dalton and Emerson Streets. It was designed in 1931 by E.A. Williams, a local architect, for the Napier Brewery Coy Ltd and it is a fine example of Art Deco in its adornment. Of two storeys and reinforced concrete, it has a splayed corner and along Dalton Street, above the main entrance, a balcony with two columns gives emphasis to the facade. This has splay arches as do the upper window heads containing sunburst motifs. The leadlights and metalwork continue the Art Deco treatment. It was (and still is) a building to gain the public's appreciation with its splendid interior reflecting a sense of wellbeing. As a schoolboy staying there en route to boarding school in the 1930s I had the pleasure of experiencing the excitement of this innovative design.

The former Napier Central Fire Station was built in 1921 in Tennyson Street, being mainly of brick and steel with reinforced concrete lintels and floor slabs. It suffered severely in the 1931 earthquake, necessitating considerable demolition of parts and rebuilding. The main two-storey portion was rebuilt in reinforced concrete and the east and west wings, which received much less damage, were strengthened. This was carried out using the practice developed for other buildings of cutting in a cage of piers and beams. The work was done in the immediate post-earthquake period and in 1960 it was refurbished as professional offices. Adjoining are the former firemen's flats and the chief fire officer's residence similarly converted to offices. Now known as the Desco Centre, it is a good example of the virtues and versatility of reinforced concrete for strengthening earlier forms of construction.

An Art Deco feature that aroused much public appreciation in its early years is the now disused Blue Baths building at Rotorua. Sited in the spacious grounds of the former Government Bath House, now Tudor Towers, it is sufficiently apart to have its own identity. The Government Architect of the day, John T. Mair, was asked to design this replacement Blue Baths building in 1929. However, when it opened in 1931 only the junior pool was ready and it was 1933 before the main pool was in use. The interior of the Blue Baths has a Roman Doric colonnade and the diving platforms are supported on modified Corinthian columns. The interplay of these classical forms with the motifs of Art Deco make an interesting combination. Blue tiles in the pools gave a splash of colour.

The general atmosphere produced by the design was that of being thoroughly modern. As a young lad on holidays in the area in the mid-1930s I always considered it a treat to swim in the Blue Baths where there was the added attraction of poolside refreshments. There was no doubt that the government, after years of neglecting the spa buildings in Rotorua, had decided to do something positive to bring attention to the importance of this international spa. The Blue Baths were noteworthy as a change of policy. They were designed for swimming rather than taking the waters, and instead of segregated nude bathing the new pools catered for mixed costumed bathing.

Blue Baths, Rotorua. This impressive Art Deco building was designed by J.T. Mair, Government Architect, as a public amenity in the Government Gardens. It is in marked contrast to the Elizabethan-style former Government Bath House. Opening in 1931 with only the junior pool available, the baths were not complete with the main pool until 1933.

A new technical development for reinforced concrete construction in New Zealand took place on a relatively large scale in the design of the water tower in Ruakiwi Road, Hamilton. The designer and project engineer, James Reginald Baird of the Hamilton Borough Council, used electric arc welding for the reinforcement in the concrete instead of laps and hoops tied with wire. At the time this was understood to have been the biggest all-welded reinforced concrete structure in the Southern Hemisphere. The tower was completed in February 1932 and the Borough Engineer, Rupert Worley, was to become one of the best known of New Zealand's eminent engineering consultants. J.R. Baird was later appointed Borough Engineer and then City Engineer. They presented a joint paper on the techniques used for this project at the International Congress of Bridge and Structural Engineering in Paris in 1932.

This circular water tower has a total capacity of 11,849,600 litres, a diameter of 25.6 m and a depth inside of 18.6 m. It consists of a reinforced concrete wall with twenty columns, Doric in style and 13.4 m high, supporting a pilastered 'frieze'. These columns are hollow and carry the overflow of water. Inside this concrete shell is a steel lining of Armco iron plate, 6 mm in thickness and welded to T-shaped iron frames to act as formwork for the concrete. The roof is of steel trusses with purlins and proprietary metal roofing which was carried out in 1973.

The contractors, W. McFadden and Son, had a base in Christchurch as well as a branch in Hamilton. Great care was taken to ensure quality control of all welding. A particular technique was evolved to achieve resistance to earthquakes for a major public utility supplying the Waikato Public Hospital as well as the central business area. Instead of lapping steel bars by as much as 1.2 m and tying with wire, the welding reduced the laps to only 76 mm. This design achieved at least a 20 percent saving in cost. The tower, sited in a park amid mature trees, is aesthetically pleasing both in its proportions and in its scale for street and park environment.

One of the most attractive hydro-electric power schemes in the country is on the Arnold River at Kaimata between Stillwater and Lake Brunner, operated originally by the Grey Electric Power Board. A feasibility study was done in 1923 but

SEA WALL, LYALL BAY, WELLINGTON.
With the bay facing Cook Strait this wall was built as a protection for Lyall Parade against rough seas and blowing sand. Designed by the city engineer's department it was completed in 1932 and has a convex shape to counter the elements.

the steam-powered station at Dobson, which opened in April 1926, caused a postponement until the next decade. Hugh Vickerman of Vickerman and Lancaster, consulting engineers, designed the Arnold River project, which was commissioned for operation on 21 September 1932. The scheme consists of a 70-m-long concrete gravity dam having a spillway over this length. On one side the dam has a concrete core earth wall. The head could be increased by the addition of collapsible crest gates on the spillway.

The water is taken by a concrete-lined tunnel of 10 sq. m sectional area from the intake through papa rock, after which it is conveyed by a concrete pipe of similar section. It crosses a river on a narrow reinforced concrete bridge, which is grassed on top as a footpath, and then enters the cylindrical concrete surge tank alongside the powerhouse. This is also of reinforced concrete, with the tailrace flowing into the Arnold River.

What makes the scheme such a visual delight is the manner in which it is integrated into the native bush. A public walkway begins at the pipe bridge and climbs to give a dramatic view over the dam before returning through bush to the pipe bridge. There is a public information kiosk as well.

As a sea-girt country New Zealand has a need for sea walls in some areas as protection against the erosive powers of the ocean. This is usually in urban locations where property is at risk. Wellington's Lyall Bay, popular for swimming and surfing, and with several lifesaving clubhouses, needed a suitably robust wall and also a promenade against occasional raging storms. The City Engineer's Department designed, in early 1931, an interesting wall which was completed in the following year. It relies for its strength on bulk and form, having a convex or ovolo moulded seaward surface to counter wave action. It also tends to lessen the amount of sand being blown across the road. This wall of considerable length has 12-mm-diameter rods at 300-mm centres vertically, with some 12-mm rods as longitudinal reinforcement. It is punctuated by a series of openings for steps with solid balustrades. As designed these steps had splendid octagonal, tapering, concrete lamp standards surmounted by the most elegant glass lamps.

Some years ago another wall was added, giving extensions to both ends. Built of proprietary concrete blocks on a rather

RETAINING WALL, UPLAND ROAD, KELBURN, WELLINGTON.
The topography of Wellington requires numerous retaining walls for streets in the hill suburbs. Such walls are often curved and add interest to the streetscape. This example was designed in 1933 and uses formwork patterns as subdued decoration.

meagre footing, it collapsed in a storm. In 1985 the original main wall subsided in one area because of undermining during a storm of hurricane intensity. Fortunately the damaged portions have been righted and repaired to sit on a new foundation, thus continuing the pleasing line of this gently curving sea wall, which has the added advantage of providing a lengthy windbreak from the northerlies for beach users sunbathing or sitting in its lee. An identically designed sea wall was built at Island Bay in 1934 and this, too, is still in use though it is starting to fracture in places.

One of Wellington's more impressive retaining walls was designed in 1933 by the City Engineer's Department for Upland Road. Near the Kelburn Viaduct, which had been built in reinforced concrete two years earlier, the wall is a massive curving structure with a slight batter and has two features which enhance its interest. Access to the houses above is provided by two curving ramps, bifurcating at street level but integrated into the total structure by solid balustrades. To provide lighting over the footpath there are two concrete lamp brackets, alas no longer with their dependent glass spheres. These brackets jut out from vertical 'pilasters', forming subtle shadow effects. A deep shadow line is also induced by the pronounced grooves which emphasise the capping to the wall both horizontally and by the ramps. The Upland Road wall is a splendid piece of streetscape, no doubt taken for granted by the hurrying motorist.

Dam, Arnold Power Station, Kaimata.

This small West Coast hydro-electric power scheme was designed by H. Vickerman of Vickerman and Lancaster of Wellington. The dam and spillway is on the Arnold River and the station came into service in September 1932 for the Grey Electric Power Board.

Opposite: Water tower, Ruakiwi Road, Hamilton.

The designer was J.R. Baird under R. Worley, Borough Engineer. Construction was completed in 1932 and it was notable for using electric arc welding for the steel reinforcement. It was the largest such work in the Southern Hemisphere at the time. The reservoir is pleasantly sited on the edge of a park.

TUAKAU BRIDGE, WAIKATO RIVER. This fine bridge was built in 1933, being the first bowstring arch example by Jones and Adams. Here the arches are higher, curves are repeated where the footway cantilevers at the arch junctions, and the balustrades are metal. With seven graceful arches, this bridge spans the river in a pleasant rural setting.

BRIDGE, MANGATAINOKA RIVER, PAHIATUA. One of several concrete bowstring arch bridges built in the 1930s, this example was opened in December 1934. It was designed by Seaton, Sladden and Pavitt, consulting engineers. Like the earlier Opawa River Bridge in Blenheim, it has low arches not seen in the later bridges.

OPPOSITE: BALCLUTHA BRIDGE, CLUTHA RIVER. This majestic structure was designed by W.L. Newnham of the Public Works Department and was opened in 1935 with great rejoicing. Six bowstring arch spans articulate this town bridge.

AOTEA CANOE MEMORIAL, PATEA.
The canoe *Aotea* is said to have brought Turi and other Polynesians from Hawaiki to Kawhia. They then came overland to Patea. This all-concrete memorial is in the main street and is of considerable interest to visitors.

Maori have long appreciated the merits of new materials and have not disdained the use of concrete. In the main street of Patea there is a striking monument which attracts visitors to the town. Constructed in 1933 of concrete throughout, it is a model of the *Aotea* canoe robustly supported on a beam with short columns. Ten seated figures in the canoe represent the captain Turi, his wife, two children, brother and five others. The *Aotea* is said to have brought Polynesian people to this district in the fourteenth century. Having a length of 16.8 m, the reproduction is impressive in scale with careful sculpting giving a lifelike semblance in an unusual but dramatic memorial.

During the 1930s some bridge engineers seem to have had a love affair with the concrete bowstring arch. One similar to the very first New Zealand example over the Opawa River in Blenheim was opened in December 1934 at Pahiatua. It consisted of rather low arches but here the diagonal braces were eliminated, never to return in this form of construction. Spanning the Mangatainoka River on the edge of the town it has seven arches of 20.7 m and was designed by Seaton, Sladden and Pavitt, consulting engineers for the Pahiatua Borough Council and the Pahiatua County Council as joint clients. The contractor was the Fletcher Construction Company Ltd. An interesting detail of this bridge is the refined shaping of the intersecting beams where the deck sits on the circular piers.

In 1933 the Waikato River was spanned by a fine bridge near Tuakau using the same principle. The Ministry of Works and Development bridge register listed an overall length of 210.8 m made up of a shore span of 11 m with two main spans of 32.9 m and four spans of 33.5 m. An unusual feature of this bridge is the inclusion at the junction of each span of a curved pedestrian bay cantilevered out over the river. The balustrades are of metal throughout. The Tuakau Bridge is most attractive in its rural setting, being uncluttered by any other structures. The designers were Jones and Adams of Auckland for the Franklin County Council.

These consultants also designed a small bridge known as Ryburn's near the settlement of Clevedon in Manukau City. Built in 1934 it has the same floriated balustrade detail as used on the Ngamuwahine Bridge four years earlier. Of single span it has deep beams which now cause an obstruction during flooding so that its days of service are numbered; a replacement bridge with a higher clearance is contemplated by the city council.

Makarau River Bridge.
Located a few kilometres northeast of Helensville, this is one of the smaller bowstring arch bridges to be found in the countryside. It was designed by the Public Works Department and completed in 1935.

Mangapurua Bridge.
Originally referred to as Morgan's Bridge, and more popularly as the 'Bridge to Nowhere' this structure has no road access. The settlers in this wild area had abandoned their farms before the bridge was completed in 1936. It is a deck arch bridge 38 m above the river. Today it is visited by tourists on foot.

In the same year a splendidly designed bowstring arch bridge was built at Balclutha over the Clutha River. Designed by William Langston Newnham, Chief Designing Engineer of the Public Works Department and later Engineer-in-Chief, it shows an appreciation of the monumentality of the arch when finely proportioned. There are six spans, each measuring 36.57 m, to give a sense of articulation and smooth transition across this waterway. It is certainly one of the best structures in the town.

The Public Works Department designed and built a small concrete bowstring arch bridge over the Makarau River in 1935. Located on a winding road linking with Warkworth, this bridge is a few kilometres north of Kaukapakapa in lower Northland. It has approach spans of 9.4 m and an arch span of 25 m to give a pleasing design in a rural setting.

Although it will have been demolished by the time this book is published, the Fitzherbert Bridge over the Manawatu River in Palmerston North is worthy of mention. Built in 1935 it had four spans of 33.5 m and seven spans of 39 m, giving a graceful structure which was rather brutally modified when a cyclists' track was hung on the upstream side. Structural problems were said to be the reason for its removal and a replacement bridge was required for the greatly increased traffic. However, with the advancement of engineering science in the structural field, one wonders whether sufficient thought was given to its stabilisation and retention as an auxiliary bridge. Palmerston North, unlike many other cities, is not so well endowed with historic monuments that it can be prodigal.

At the end of the First World War some returned sevicemen were settled on rural land under a scheme arising from the Discharged Soldiers Settlement Act 1915. Not all land selected was really suitable, one such area being the remote steep hill country in the Mangapurua district on the true left bank of the Whanganui River north of Pipiriki. It was reached by the long

Ryburn's Bridge, near Clevedon.
Opened in 1934 this is the work of Jones and Adams, forerunners of today's KRTA Ltd. A hallmark of their work was the floriated pattern in the solid balustrade to give a less daunting effect.

Opposite: Omakau Bridge, Manuherikia River.
Omakau in Central Otago is linked to Ophir by this bridge, erected in 1938. It was designed by Vickerman and Lancaster for the Vincent County Council and is rather unusual in having splayed concrete solid balustrades. It reminds one of the flumes and water races of the region.

WAIAU STREAM VIADUCT, KOPUAWHARA VALLEY.
This dramatic railway structure on the Napier-Gisborne line was designed in 1939, but not completed until 1942 because of wartime shortages of labour and materials. It is a superb bridge by any standards and incorporates innovative techniques for earthquake resistance.

TANGAHOE VALLEY BRIDGE.
This structure, about 15 km northeast of Hawera, is a bowstring half-arch bridge — one having the deck passing through the arch. It replaced a bridge destroyed by floods and was built in 1937 to a Public Works Department design.

Opposite: FAIRFIELD BRIDGE, WAIKATO RIVER, HAMILTON.
Another of the many bridges designed by Jones and Adams is this fine example built in 1937. Here attention has been given to a smooth transition between the lower ends of the bowstring arches. The curve has been used for the footways as they sweep round to meet the streets on both sides of the river. This bridge shows refinement in a suburban setting.

MILLER'S BUILDING, TUAM STREET, CHRISTCHURCH.
Now the civic offices of the Christchurch City Council this building is of flat slab construction with mushroom column caps. Refurbishment from a large department store to offices caused these to be hidden in suspended ceilings. It was designed by G.A. Hart and built in 1938.

winding Mangapurua Road which continued from the Manganui-o-te-Ao Valley to the north of Raetihi. A suspension bridge across the Mangapurua Gorge had been erected in 1919 but this became dangerous through rotting. A design for its replacement was prepared by the Public Works Department in 1933 and eventually tenders were called for its construction on a labour-only basis. The successful tenderer, Sandford and Brown of Raetihi, began work in January 1935. The bridge was to be of reinforced concrete having an arch of 23 m, a total length of 39.6 m and a deck height above the water of 38 m.

Extremely wet weather made access roads impassable and created such a problem for the contractor that no work could be done for five months. The situation was aggravated by the late supply of materials, the responsibility of the PWD, and so the termination of the contract was requested, but to no avail. Eventually the work was done; the official completion date was 5 June 1936 and no penalties were enforced. However, the irony was that many settlers had left the area before the bridge was finished. They had abandoned their farms as being totally uneconomic, the acute isolation and the Depression years making life there intolerable. The new bridge soon became known as the 'Bridge to Nowhere' and so it has remained for many years, with no usable vehicle access. A solitary monument swallowed up by regenerating bush it is seen only by visiting trampers.

A small bowstring arch bridge in a rural setting is in the Tangahoe Valley east of Hawera. It has a single-arch span of 24.23 m with two small approach spans of 2.8 m, the deck being 11.58 m above normal water level. This bridge was built in 1937 to replace one destroyed in major floods in the previous year. An interesting feature is that the deck passes through the arch at about midpoint, the legs of the arch being anchored into the abutments in a visually impressive manner. The Tangahoe River Bridge was designed for the Hawera County Council by the Public Works Department to give a 2.4-m clearance above the highest recorded flood level.

As we have seen, concrete bridges can take a variety of forms and very many older examples are still in use. In the late

1920s the Vincent County Council engaged the well-known Wellington firm of Vickerman and Lancaster as its engineering consultants for an extensive programme of bridge building. One of the bridges, at Omakau over the Manuherikia River, replaced a footbridge giving access to Ophir. The new bridge, designed in 1937, opened in the following year. Looking very much like a large-scale flume or water race, it has splayed solid concrete balustrades from its narrow one-way deck. The foundations consist of three octagonal piles per pier to take the haunched reinforced concrete beams, which are monolithic with the deck. There are ten spans, two of 3 m and eight of 11.5 m, giving a total length of 98 m.

SCHOOL DENTAL CLINIC, WILLIS STREET, WELLINGTON.
This building incorporates large shear walls believed to have been the first over 18 m in height when erected in 1939. The design was from the Government Architect's Office where structural engineers were involved from the outset.

For some the epitome of the concrete bowstring arch is seen in the Fairfield Bridge, built in 1937 over the Waikato River in Hamilton. It was designed by Stanley Jones of Jones and Adams for the Hamilton Borough Council, with the neighbouring county councils also contributing funds towards its cost. There is one span of 39.6 m and two of 39 m, with the two end spans being 10.6 m to give a total length of 139.3 m. Of all the concrete bowstring arch bridges this structure shows a rare refinement in the manner in which the lower ends sweep into each other with a smooth transition. The subtle exploitation of the arch is further seen in the curved footpath approaches. The balustrades follow the established practice of the designer in having a pierced pattern — in this bridge, geometric rather than floriated. The encroachment of houses along the banks has not diminished the setting unduly, which is still attractive with trees and shrubs. This is certainly a bridge to be admired. Its distinctive form provides a contrast to the splendid steel arch Victoria Bridge some distance upstream.

One of the later essays in flat slab construction is the former Miller's Building in Tuam Street, Christchurch. It was built in 1938 to the design of George A. Hart and embodies the expression of the clean, uncluttered envelope as a desirable modern attribute. A large structure of five floors, it has been considerably changed from its use as a department store to the civic offices of the Christchurch City Council. As such it has been handled splendidly, although for some it is unfortunate that the mushroom column caps have been hidden in the suspended ceilings. Furthermore, on the ground floor the octagonal column shafts have been screened by circular sleeves. As a contribution to the streetscape it is a most worthy building.

The honour of being the longest bridge in New Zealand goes to the Rakaia Bridge, opened in 1939 on State Highway 1 near the town of the same name. It is not visually exciting, being a ribbon-like structure of relatively small spans and low level. It has a concrete deck on beams and the standard late 1930s type of pierced balustrade having slots with rounded ends. The 143 spans of 12.9 m give a total length of 1.75 km. The deck is 7.3 m wide.

Throughout the 1930s and thereafter into the postwar years, the Public Works Department's Architectural Branch developed a close association with structural engineers who became an integral part of that office. This provided rewarding experience for both engineering and architectural disciplines in allowing the structure to be considered from the earliest design concept, and in providing continuity of collaboration with mutual respect.

The School Dental Clinic building in Willis Street, Wellington, was built in 1939 and is believed to be New Zealand's first reinforced concrete shear wall building over 18 m in height. Designed as a prototype it has transverse shear walls with deep longitudinal walls pierced for openings. This building is the training centre for the school dental service. As such its main floor is given over to full-length surgery space for rows of chairs for the trainees and their small patients. The 1942 earthquake of 7+ on the Richter Scale and centred in the Wairarapa had no discernible effect on the building.

Four identical structures broke new ground in concrete technology when they were completed in 1939. They were the hangars at Ohakea and Whenuapai Air Force Stations. The main buildings erected for the establishment of these bases were of reinforced concrete and included an officers' mess, sergeants' mess and recreation hall.

The hangars, two at each station, were designed by R.G. Caigou under the direction of Charles William Okey Turner, Chief Designing Engineer of the Public Works Department. The working drawings are dated December 1937. The huge arched roofs have a total span of 82.9 m and consist of 0.6-m-deep ribs with a 100-mm-thick slab. The ribs were anchored outside the hangar building proper, which has 8-m-wide workshops and ancillary rooms at the sides and larger ones along the rear. Whether or not this design was inspired by the work of Pier Luigi Nervi, the brilliant Italian engineer, it certainly reflected the growing maturity of structural design in New Zealand at this period.

The last highway bridge to get a mention in this book was designed in 1939 but not opened for traffic until 1940. The Fish River Bridge spans a deep gorge stream on the Otago side of Haast Pass. The design was carried out by Jack Hanlon, at that time on the staff of the PWD. He later became a well-known structural engineering consultant in Dunedin and widely respected for his competence and forward thinking.

HANGAR, OHAKEA AIR FORCE STATION.
This is one of four identical hangars completed in 1939: two at Ohakea and two at Whenuapai. Designed under the direction of C.W.O. Turner, Chief Designing Engineer of the Public Works Department, they have a span of 83 m and the slab between the ribs is only 100 mm thick.

FISH RIVER BRIDGE, HAAST PASS.
This attractive bridge is rarely seen by motorists who speed over it without any idea of its beauty. Designed by J. Hanlon of the Public Works Department in 1939, it was opened in 1940.

This bridge has a concrete arch of 31 m span, which cannot be appreciated with the straight alignment unless one climbs down to the streambed. There are also two landing spans of 4.9 m and 5.2 m to make it a very attractive structure in its setting of beech forest.

We conclude this survey of concrete structures with a dramatic rail bridge on the Napier–Gisborne line. The last of the six viaducts to be built on the 191-km railway was the Waiau Stream Viaduct designed in 1939. A design for one in steel with plate girders on trestle piers was compared with a reinforced concrete arch and girder approach spans. One of the deciding factors in the choice was the wartime shortage of structural steel, and the concrete design proceeded.

The result is a superb structure 162 m long, having an open spandrel parabolic arch of 54.9 m span and a height of 30.5 m. There are five reinforced concrete girder spans of 12.2 m and one of 9.1 m on the south side, and three 12.2 m girder spans on the north — all supported on reinforced concrete trestle piers. When completed in 1942 after wartime delays of labour and materials, it was rightly hailed as a splendid design. It has completely monolithic construction and the abutments have Mesnager hinges. Longitudinal forces are taken by compression in the arch ribs and lateral forces by bending of the bridge deck and transmission to the main bents. Visually it is a delight.

Conclusion
Utility and diversity

While this survey has covered the period from the 1850s to 1939, it is not an exhaustive record by any means and there will be other structures still in existence. Because of the steady increase in buildings and bridges erected from the early twentieth century in reinforced concrete (commonly called ferro-concrete in the first two decades), I have been selective so as to give a fair representation for this group.

What has become obvious as a result of my research is the number and variety of plain concrete structures built in the nineteenth century. It is worth considering again the probable reasons why a small, newly established British colony made such early use of a new building material.

Because the colonisation of New Zealand was an ongoing process for much of the Victorian period, new technology was brought in by some immigrants from time to time. In general the traditional nature of British building technology, with its heritage of skilled trades such as masonry and bricklaying, militated against concrete usage for all but strictly utilitarian construction. Although there was a measure of early innovative concrete work in England, patents and practical attempts at building were shortlived. Indeed concrete was seen as more suitable for exterior plasters for dressings, mouldings and general decorative use in classical motifs. It was also favoured for precast units such as garden ornaments, urns and balustrading. Seldom was it used in entire buildings.

Here in New Zealand there was a strong motivation by many settlers to become more adventurous. This was already evidenced in their decision to uproot from their homeland and travel 19,000 km to re-establish themselves in the Antipodes. They soon realised that skilled masons and bricklayers were not always available, especially in rural areas, nor were supplies of suitable stone and brick. Timber, as the predominant building material, was often regarded as temporary and inferior, so they looked for something more substantial. Earth-building, such as cob and adobe, was quite widespread in the drier regions in the 1860s and seventies, and it is likely that the techniques for concrete building were considered as generally similar to those for earth-building. Cement was imported from England from the early 1840s and it was not too difficult to locate supplies of sand and gravels for making concrete. Moreover it seemed that a do-it-yourself approach was appropriate where the cost could be reduced by providing one's own unskilled labour. New Zealand established its first cement works in 1884, ahead of Australia.

Probably a cogent reason for concrete usage was the prevalence of destruction of timber buildings by fire, especially in areas where water supplies, firefighters and appliances were not available or were inadequate. This was in fact the position generally throughout the nineteenth century. It is also easy to appreciate the civil engineer's interest in concrete. Here was a material ideal for bridge piers and abutments, foundations, breakwaters, other harbour works, fortifications and so on.

Many engineers and architects have been mentioned throughout these chapters. In the nineteenth century they were born and trained predominantly in Britain, and they had to adapt to the restrictions and challenges of their professions in a raw colony. Many did so by adopting a pragmatic approach, learning from failures, and building up a store of expertise. The next generation gained immeasurably from this body of knowledge, with resultant benefits to the community.

The advent of reinforced concrete was a real boon to New Zealand engineers and architects. Early attempts in the late 1870s and early eighties, although advanced at the time, were rather primitive and lacked the benefit of proper design. Reinforced concrete meant that much larger structures could be built. Moreover, resistance to earthquakes and wind forces could be incorporated. Multi-storey buildings sprang up from 1906 onwards, as well as bridges built as arches or as beam structures supported on piers and abutments. Concrete water towers were more practicable and concrete dams were to become more reliable than those of earth or stone.

An understanding of the plasticity of concrete opened up many new aspects in design and construction of reinforced concrete structures, for example in hangar roofs where thin shells and ribs gave much greater clear spans, and in surfaces of double curvature such as turbine scroll cases for hydro-electric stations.

Although earthquake-resistant design and construction was not properly understood in this early period, some important pioneer work was carried out by C.R. Ford, S. Irwin Crookes and his sometime engineering partner Siacci in the 1920s. The disastrous Hawke's Bay earthquake of 3 February 1931 resulted in an effective building code requiring proper seismic design, especially for built-up areas.

There was a steady improvement in the quality of concrete after the first decade. Hand placement and tamping around steel reinforcement required care and it was necessary to ensure adequate daily supervision by inspectors and clerks of works. Later, the use of mechanical vibrators produced speed and efficiency. Furthermore, the design of concrete mixes improved with a greater understanding of the aggregates and the availability of high grade cements. From the 1930s there was a greater adherence to national standard specifications.

It has not been possible to accurately correlate the amount of concrete building in New Zealand in the nineteenth century with that in other countries. However, from the number of surviving examples it certainly appears to have been more widely adopted here, considering our very small population, than in any other country. New Zealand did rather well.

Bibliography

Official publications

Appendices to the Journals of the House of Representatives: Public Works Statements.
Blue Books 1843–47.
Handbook of New Zealand Mines 1887.

Theses and unpublished manuscripts

McEwan, Ann E. 'From cottage to "skyscrapers". The architecture of A.E. & E.S. Luttrell in Tasmania and New Zealand'. MA thesis, University of Canterbury, 1988.
Hamilton, Judith. 'Sunnyside Hospital'. Sub-thesis, n.d.
Shanahan, Kieran J. 'The Work of William H. Gummer'. B. Arch. thesis, University of Auckland, 1983.
Wilson, P.R. 'The Architecture of Samuel Charles Farr 1827–1918'. MA thesis, University of Canterbury, 1982.
Wilson, G.B. 'History of Wilson's Portland Cement Coy'. N.p. n.d.
Wilson, T.H. 'History of Wilson's (NZ) Portland Cement Ltd 1884–1956 and Reminiscences of T.H.Wilson'. N.p. n.d.

Booklets, pamphlets, papers and journals

Coutts, Peter J.F. *Towards the Development of Colonial Archaeology in New Zealand*. Part 1. 1983.
Lewis, M. *Monier and Anti-Monier: Early Reinforced Concrete in Australia*. Second National Conference on Engineering Heritage. Melbourne, 1985.
Lippincott, R.A. 'The Development of Concrete as an Artistic Architectural Material.' *The Journal of the New Zealand Institute of Architects*. Wellington, 1928.
Macdonald, C. Fleming. 'Reinforced Concrete.' *Progress*, April 1913.
Mackenzie, Catherine (ed). *Centenary of Barrhill 1877–1977*. N.p. n.d.
New Zealand Concrete Research Association. *The Manufacture of Portland Cement*, Information Bulletin, November 1984.
New Zealand Electricity Department. *Electric Power Transmission*. Government Printer, Wellington, 1960.
'No. 5 Pumphouse Martha Mine Waihi.' *Ohinemuri Regional History Journal*, 1971.
Progress — Incorporating The Scientific New Zealander, 1905–12.
Reynolds, I.B. *Cast in Concrete — The Substance of New Zealand Building*. Hopkins Lecture, IPENZ, 1982.

Stanley, Christopher C. *Highlights in the History of Concrete*. Slough, 1980.
Wilson's Portland Cement Co. Ltd. *The Roadway of the Future: A Few Facts about Concrete Roads*. Auckland, 1916.

Reports
'AMP Building', New Zealand Historic Places Trust.
'Arden', NZHPT.
Courtville', NZHPT.
Cochran, C.C. 'Rimutaka Railway Structures.' Report for NZHPT.
Matheson, Ian. 'Rangitane Toll Bridge & Tane Flaxmill.' Report for Manawatu Regional Committee, NZHPT.
Norris, H.C. 'Ngaruawahia Flour Mill Store.' Report for NZHPT.

Encyclopaedias and works of reference
Brees, S.C. *The Illustrated Glossary of Practical Architecture and Civil Engineering*. London, 1853.
Cyclopedia of New Zealand. 6 volumes. Wellington & Christchurch, 1897–1908.
Guedes, Pedro (ed). *The Macmillan Encyclopedia of Architecture and Technological Change*. London, 1979.

Books
Bates, Arthur P. *Bridge to Nowhere*. Wanganui, 1981.
Bill, Max. *Robert Maillart*. Zurich, 1949.
Blair, W.N. *The Building Materials of Otago*. Dunedin, 1879.
Bowman, H.O. *Port Chalmers: Gateway to Otago*. Dunedin, 1948.
Carney, Samuel A. *Mill Hill's 100 Years The Story of St Joseph's Missionary Society 1866–1966: The Years in New Zealand*. Putaruru, 1966.
Collins, Peter. *The Vision of a New Architecture: A Study of Auguste Perret and his precursors*. London, 1959.
Concrete Publications Ltd. *Concrete Roads and Their Construction*. London, 1920.
Fearnley, Charles. *Early Wellington Churches*. Wellington, 1977.
Francis, A.J. *The Cement Industry 1796–1914: A History*. London, 1977.

Furkert, F.W. *Early New Zealand Engineers*. Revised and edited by W.L. Newnham. Wellington, 1953.

Gambrill, M.D. *A History of Queen Margaret College*. Wellington, 1969.

Gee, David. *The Devil's Own Brigade: A History of the Lyttelton Gaol 1860–1920*. Wellington, 1975.

Gordon, Mona. *The Golden Age of Josiah Clifton Firth*. Christchurch, 1963.

Hawkins, D.N. *Rangiora: The Passing Years and People in a Canterbury Town*. Christchurch, 1963.

Hawkins, D.N. *Beyond the Waimakariri: A Regional History*. Christchurch, 1957.

Hering, Oswald C. *Concrete and Stucco Houses*. New York, 1922.

Keys, H.J. *Mahurangi: The Story of Warkworth*, New Zealand. Warkworth, 1953.

Lemon, Daphne. *Taieri Buildings*. Dunedin, 1970.

Lemon, Daphne. *More Taieri Buildings*. Dunedin, 1972.

McAra, J.B. *Gold Mining at Waihi 1878–1952*. Christchurch, 1978.

McDonald, K.C. *White Stone Country: The Story of North Otago*. Dunedin, 1962.

McDonald, K.C. *City of Dunedin: A Century of Civic Enterprise*. Dunedin, 1965.

McKay, J.G. and Allan, H.F. (eds) *The Nelson Old Boys' Register 1856–1956*. Nelson, 1956.

Millar, F.W.G. *Golden Days of Lake County*. Christchurch, 1973.

Morrison, Robin. *Images of a House*. Martinborough, 1978.

Musgrove, S. (ed) *The Hundred of Devonport: A Centennial History*. Devonport, 1986.

Napier, W.J. *Notes on Harbours and Docks with special reference to reinforced concrete*. Auckland, 1907.

Rockel, Ian. *Taking the Waters*. Wellington, 1986.

Rose, John. *Akarana, Port of Auckland*. Auckland, 1971.

Ross, John O'C. *The Lighthouses of New Zealand*. Palmerston North, 1975.

Rolt, L.T C. *Victorian Engineering*. Harmondsworth, 1974.

Sherrard, J.M. *Kaikoura: A History of the District*. Kaikoura, 1966.

Stacpoole, John Bevan Peter. *Architecture 1820–1970*. Wellington, 1972.

Stewart, Peter J. *Days of Fortune: A History of Port Chalmers* 1848–1973. Dunedin, 1973.

Index

Abbeystead Dam, England 100
Abbotsford *28*, 62
Adams, E.F. 174
Adams, William 28
Addington, Christchurch 66, *67*, 69
Admiral Codrington 149
Admiralty Jetty, Auckland 112
Aggregate 8, 10, 12
Akers, Hugh 159
Alexandra 69
Alves, John 39
Amberley 77, 84, *85*
AMP Building, Wellington *200*, 202
Anglican Cathedral, Christchurch 24
Anzac Bridge, Kaiparoro 172, *184*
Aotea Canoe Memorial, Patea 222, *222*
Appleby House, Te Awa 75
Arapuni Dam 202, *208*
Arapuni Powerhouse 203
Arden 192, *193*
Ardross 76, 77, *85*
Arkwright, Francis 73, 83
Armstrong Breech Loading Gun 75
Arnold River 215, 216, *219*
Arnold Hydro-electric Power Station 215, *219*
Art Deco 196, 213, 214
Arts and Crafts Movement 53, 132
Arthurs Point Bridge (Edith Cavell) 159, *160*
Ashley River Bridge 130, 142, *143*
Aspdin, Joseph 13, 14
Aspdin, James 14
Aspdin, William 14, 16
Aspdin & Coy 14
Atkins, Alfred 73
Atkins and Bacon 115
Auchmore, Taieri 92, *93*
Auckland 26, *27*, 32,*32*, 35, 70, *70*, 71, *71*, 72, 74, 76, *78*, 80,*81*, *84*, 105, 118, 124, *125*, *126*, 128, *138*, 145, *161*, 162, 174, 176, 177, *180*, 187, 188, 190, 194
Auckland City Council 71, 125
Auckland Gas Company 87
Auckland Master Builders' Association 128
Auckland Harbour Board 77, 78, *180*, 187
Auckland Railway Station 210, *211*
Auckland Technical College (ATI) 130, *131*
Australian Mutual Provident Society (AMP), Wellington *200*, 202
Australian Telefunken Company 136

Baird, James R. 215
Balclutha 40, 64
Balclutha Town Bridge 220, *221*, 223
Balfour, James M. 156, *157*
Balgownie, Naenae, Hutt Valley 94
Bankside Street Cottage, Auckland 70, *71*
Barnhill, Thomas 62
Barnicoat, Hon. J.W. 210
Barnicoat House, Nelson College, Nelson 211, 212
Barrhill, Mid-Canterbury 54, 59, 60, *61*

Barrhill School 54, 61
Bartley, Edward 72, 80
Basham, Frederick 147
Basin Reserve Pavilion, Wellington 188, *189*
Basset & Co, Wanganui 130
Bastia Hill, Wanganui 194, *199*
Baudot, Anatole 17
Bay of Plenty 135, *137*, 210
B & M Cyanide Tanks 93, 112, 113, *119*, 120
Bedford Road, Taranaki 103, 105
Bell, Arthur, W.D. 75
Bell, Charles N. 72, 79
Belmount, Hawke's Bay 162, *165*
Berkeley, Taieri 62
Bertinshaw, G.W. 117
Beverley, William 48
Blackett, John 71, 72, 157
Blacksmith's Shop (smithy) *37*, *40*, 62, 63
Blackwater Dam, Scotland 100
Blair, Matthew 124
Blair, William N. 123
Blechynden, J. 154, 155
Blenheim 55, 132, *132*, 158, 159, 166, 167
Blenheim Post Office 55
Blomfield, Sir Arthur 10
Blue Baths, Rotorua 214, *215*
Blue Cliffs station, South Canterbury 42
Bluff 156
Boddam, Major Edward T. 75
Bournemouth, England 18, 69
Bowmar House, Tara 76, *77*
Bowmar, Joseph 76
Bowstring arch bridges 158, 159, *167*, *220*, *221*, 222, *223*, *226*, *227*, 228, 229
Boxing 9
Bradley's Hop Kiln, Dovedale 73
Breweries 54, *212*, 214
'Bridge to Nowhere', Mangapurua *223*, 228
Bridge of Remembrance, Christchurch 178, *179*, 186
Brigham, G.W. 134
Bristol Road Bridge, Taranaki 128, *130*
Britain 13, 16, 18, 69, 112, 134, 233, 158
British Plate Glass Coy, London 13
Brogden Bros 72
Broken River 122
Brook Dam, Nelson 100, 158
Brown, C.F. 113
Brunel, Isambard K. 14
Brydone, Thomas 58
Bullen, Frederick 39
Bullen, George F. 39, 90, 91
Bullen Brothers, 40
Burgess 134
Burgess Island, Mokohinau Group 71
Bush, Walter E. 120, 174

Caigou, R.G. 230
California 18, 30
Calliope Dock, Devonport 77, *78*, 112

Cameron, C.P. 146
Cameron, J.C. 140, 146
Campbell, Alexander 24, 92
Campbell, John, Government Architect 145
Campbell, John L. 26
Campbell, Robert 24
Campbell, Robert (England) 58
Campbell, Robert jnr 58
Campbell, William 48
Canada 15
Cane, Thomas 30
Canterbury 90, 122
Canterbury Jockey Club 170
Cape Foulwind 193, *198*
Cape Kidnappers 56
Cape Terawhiti 147
Cargill, Edward B. 48
Cargill, William 48
Cargill's Castle, Dunedin *47*, 48
Cass 122
Cassell Gold Extraction Company 90
Castlamore (Woodside), Dunedin 43, *45*, 48
Castlerock station, Southland 62, *63*
Cathedral of the Blessed Sacrament, Christchurch 101, 110, *111*
Cautley, Major Henry 74, 75
Cement City, USA 176
Cement, Medina 14
Cement, Portland 9, 13, 16, 31, 71, 87, 88, 89, 102, 123, 176
Cement, Roman 13, 14, 15, 16
Central Otago 8, 98, 108, 161, 166, 173, 184, 197
Central Hotel, Napier 213
Central Fire Station, (Desco Building) Napier 213, 214
Cezanne, Louis 21
Chambers, Mason 152
Chambers, Maurice 192
Chapman, Henry, Judge 43, 48
Chapman-Taylor, James W. 132, 134, 152
Chapman, Skerrett, Wylie and Tripp Building, Wellington 116, *117*
Chicory Kiln, Inchclutha 64, *65*
China 12
Christchurch 24, *29*, 30, 42, 66, *67*, 101, *111*, 114, 117, 170, *171*, 178, *179*, *228*, 229
Church of Our Lady of Lourdes, Rotokakahi 193, *195*
Civic Offices (Miller's Building), Christchurch *228*, 229
Clark, Jack 161
Clarence Reserve station, Inland Kaikoura 39
Clayton, William H., Colonial Architect 31, 54, 55
Clere, Frederick de J. 73, *83*, 116, 117, *117*, 118, 128, *129*, *139*, 144, 168, 171, *171*, 172, *183*, *200*, 202,
Clevedon 210
Clifden Bridge, Southland 94, *95*
Clifton, Auckland 26, *27*
Climie, Henry W. 186

Clive Grange Estate and Railway Coy Ltd 56
Clutha River 64
Coignet, Francis 17
Collingwood 88
Concrete, town, USA 176
Concrete High School USA 177
Concrete blocks 92, 121, 122, 123, 124, 174
Concrete, Camerated 117, 118, 130, 156
Concrete, Ferro- 101, 102, 105, 232
Concrete, mass (plain) 9, 90, 92, 100, 232
Concrete, no-fines 39
Concrete, Orotonu 144
Concrete, prestressed 20
Concrete, pumice 114
Concrete, reinforced 16 et seq, 98 et seq.
Concrete mixers 21
Concrete roads 174, 176
Condensation 173
Cookshops 56, *57*, 58, 62, *63*
Cook Strait 140, 216
Coplay, USA 16
Corlett, B.S. 113
Corner Courtville, Auckland *161*, 162
Corwar Lodge, near Barrhill 53, 54, *59*
Courtville, Auckland *138*, 145
Craggy Range house, Hawke's Bay 162, *165*
Craigieburn, Mid-Canterbury 122
Crampton, Thomas R. 88
Crinan Street, Invercargill 124
Cromwell 172
Cromwell Gorge 156, *166*
Crookes, S.Irwin, Jnr 20, 233
Cross Creek 43
Crown Mine Battery, Karangahake 90, 91, *96*
Cull, John E.L. 156
Cyanide Tanks, Waihi *119*, 120, 122

Dalgety and Coy, Blenheim 132, *132*
Dams 20, 31, 100, 115, *116*, 128, 158, *166*, 189, 202, 203, 204, *208*, *209*
Daniell, F.C. 118, 156, 189
Davis, James E. 176
Davidson, W.S. 58
Davis Road Extension, Taranaki 103, 104,
Dawson, J. 158, *166*
Dawson, J.M. 130, *131*, 158
DB Breweries Ltd 212,
DB Central Brewery, Mangatainoka 212,
de Montalk, Robert W. 128, 162, *163*
Department of Tourism and Health Resorts 113
Desco Centre, Napier *213*, 214
Devonport 74, 77, *78*, 80
Devonport Flats, New Plymouth 188, *188*
Dobbs, Edward 13
Dochring, C.E. 19
Dodd, Ralph 16
Dog Island Lighthouse 156, 157, *157*, 158
Dominican Priory, Dunedin *44*, 49
Dominion Breweries 212
Dominion Portland Cement Company 89

237

Donald, James 24
Donovan's Store, Okarito 24
Dovedale, Nelson District 74
Drummey, Jeremiah 69
Dunback, North Otago 88
Dunedin 21, 28, 31, 32, 34, 38, 39, *44*, *45*, *47*, 65, 75, *91*, 91, 98, 99, 102, 115, *115*, 135, *142*, 144, *191*, 192, 230
Dunedin City Council 174
Durham Road, Taranaki 105, *105*
Duthie, John 94

Earth buildings 23, 24, 26, 232
Eastbourne 144
East Coast 135, 148, 182
East Cowes, England 14
Edith Cavell Bridge 159, 160, *160*, 161
Eddystone Lighthouse, England 13
Egmont Brewery, New Plymouth 54, *54*
Elderslie Estate, North Otago *11*, 39
Elizabethan Style 73, 83, 215
Ellis, Peter 66
Eltham 147
'Engineering to 1990' 113
England 9, 15, 17, 22, 26, 71, 112, 123
Errington, William, 77
Ettrick, Central Otago 172
Ewer's Hop Kiln, Upper Moutere 74, *93*, 94
Experimental Metallurgical Works, Thames 76

Fairfield Bridge, Hamilton *227*, 229
Falkner, A. 172
Farr, Samuel C. 49, 50, *50*, 51, 62, *63*, 64, *64*, 77
Featherston 24
Featherston County Council 132
Felkin, Harriet 149
Felkin, Dr Robert 149, 152
Fenceposts 122, 126, *127*
Ferntown 88
Ferro-concrete 10, 101, 102, 105, 232
Ferro-Concrete Company of Australasia 105, 118, 120
Firth, Josiah C. 26, 27, 65, 66, 86, 112
Fish River Bridge, Otago 230, 231, *231*
Fitzherbert Bridge, Palmerston North 223
Flatbush School Road Bridge, Manukau City 177, *177*
Fleming, Henderson and Coy 162
Fletcher, James 162
Fletcher Construction Company 162, 222
Ford, Charles R. 20, 233
Foret, R. 20
Forrest, Robert 42
Fort Ballance 74, *75*
'Fort Cathie' 135
Fort Cautley 74, *80*
Fort Jervois 75
Fort Taiaroa 75
Fort Takapuna 74, *80*
Fox, Dr 16
France 15, 17
Francis, Charles and Sons 14
Franklin County Council 222
Freezing Works 148, 149, *150*, *151*, 170, *182*
Friend, W.Leslie 117
Freyssinet, Eugene 20
Furkert, Frederick W. 159, 160, *160*, 188
Fyffe, George 22
Fyffe House, Kaikoura 22

Gare Maritime, Le Havre, France 20

Gateshead-on-Tyne, England 14
Geary 123
Gentle Annie Creek Bridge, Central Otago 172, *173*
George Street Bridge, Dunedin 98, *99*
Germany 15
Giaconda, Fra 13
Gisborne 148, 178, *179*, 231
Gisborne Sheepfarmers Frozen Meat Company 149
Glenmark station, North Canterbury 49, 50, 62, *63*, 64, *64*
Glue, Walter F. 189, *206*
Glue, William P. 189, *206*
Glue Construction Coy 189, *206*
Goddard, Henry A. 118
Golden Bay 88
Golden Bay Cement Company 88, 170
Golden Bay Cement Works Limited 88
Goldies Brae, Wellington *41*, 42
Gore 161, *163*
Goreham, William 15
Gothic 30, 43, 46, 49, 58, 60, 64, 114, 138, 168, 172, 189,
Government Bath House, (Tudor Towers) Rotorua *110*, 113, 114
Gow, John 23, 25
Grafton Bridge, Auckland 118, 119, 120, *125*, *126*, 161
Grain Store, St Andrews, South Canterbury *56*, 58
Grandstand, Riccarton Racecourse, Christchurch 170, *171*
Grandstand, Trentham Racecourse 170, *181*
Grandstand, Wanganui Racing Club *11*, 42
Grant, John 15
Grant, R. Allan 29, 30
Graving Dock, Port Chalmers 72, *113*, 114
Graving Dock, (Calliope) Devonport 77, *78*, 112
Graving Dock, Lyttelton 72, 79
Grayling, F.M. 105
Great North Road, Auckland 174
Great South Road, Auckland 174
Green Hayes, Temuka 73, *82*
Greenhills Run, Inland Kaikoura 39, 40
Gregg and Coy 65, *65*
Grey Electric Power Board 215, *219*
Grey Lynn, Auckland 75, *84*
Gropius, Walter 18
Grosvenor Terrace, Wellington 162, *163*
Gumbley, E.A. 194, 199
Gummer, William H. 152, 153, *153*, *162*, 192, *193*
Gummer and Ford 210, *211*, 212, *212*
Gummer and Prouse 178, *179*

Haast Pass 230. 231, *231*
Hadrian's Wall, England 12
Hamer, W. H. 105
Hamilton 118, *154*, 155, 189, 215, *218*, *227*, 229
Hamilton Borough Council 215
Hangar, Ohakea Air Force Station, Manawatu 230, *231*
Hanlon, Jack 230, 231, *231*
Hart, George A. 205, *228*, 229
Hastings 138, 144, 176
Hauraki Gulf 71
Havelock North 152, *152*, 153, *153*, 162, *165*,192, *193*
Hawera *141*, 146, 228
Hawera County Council 228

Hawkes Bay 56, 94, *100*, 101, *152*, 152, *213*, 213
Hawkes Bay Earthquake, 1931 159, 162, 196, 233
Hay, Robert 88
Hayhurst, John T.M. 73, *82*
Hector, Dr James 31
Henderson and Paul 54
Henderson, A. & K. 194, *201*
Hennebique, François 17, 18, 19, 69, 104, 105
Herapath, Philip 32, *32*, 35
Heretaunga Settlement, Petone *100*, 101
Hering, Oswell C. 173
Hicks Bay, East Coast 148, 170, *182*
Higginson, Harry P. 51, *52*
Hislop, James 88
Hobson Wharf, Auckland 187
Holmes, Mathew 62
Hornell, Professor P.G. 203
Horotiu Bridge, Waikato 162
Horsley, Norman 87
Hotel Cargen, Auckland 128
Howorth, C.W. 94
Hukarere School, Napier 130
Hunterville *204*, 205, 206
Hurry and Seaman 88
Hyatt, Thaddeus 17

I.C. Johnson and Company 14, 15
Inchclutha 64, 65, *65*
Iles, E.J. 190
India 15
Inglewood 105
Inglewood County Council 104, 105
Inland Kaikoura 40
Invercargill *120*, *121*, 123, 124
Invercargill Prison 123
Invermay, near Mosgiel 23, *25*
Island Bay, Wellington 217
Island Block, Central Otago 172, *184*

Jack Brothers 168
James, L.G. 148, 150
James, Peter 76
Jellicoe, Lord 187
Jervois, Lt.Gen. Sir William 74
Jessop, William 16
Johnson, Isaac C. 14
Johnson, Captain Robert 72
Johnston, Dr Alexander *41*, 42
Jones and Adams, 162, 177, *177*, 190, *191*, 210, *220*, 222, *224*, *227*
Journey's End run, North Canterbury 58

Kahutara, Inland Kaikoura *36*, 39, 90
Kahutara River 39
Kaikoura 22, 36, 39, 90, *91*
Kaimata, West Coast 215, *219*
Kaiparoro Memorial Bridge, Wairarapa 172, 178, *184*
Kaitaia 136, *136*
Kaiti Bridge, Gisborne 178, *179*
Kaiwarra Stream, Wellington 31
Karangahake 90, *96*
Karori 31, *33*, 115, 128, *129*, *133*, 134, 135, 144
Karori Borough Council 134
Karori Park *133*, 134
Karori Recreation Ground *133*, 134, 135
Karori Reservoir 31, *33*, 115. 116, 128
Karori Rock Light *140*, 148
Kawarau Gorge *173*
Kawarau Falls Bridge 190, *197*
Kawarau River 172, 190

Kawau Island 86
Keene Brothers 40
Keene's Cement 15
Kelburn, Wellington 217, *217*
Kennedy 123
Kereu River, Bay of Plenty 135
Khyber No 2 Reservoir, Auckland 130
Kiln, cement 15, 87, 88, 89
Kiln, chicory 65, *65*
Kiln, hop 73, 74, *93*, 94
Kiln, lime 86, 87
Kiln, timber 168, *169*
Knobby Range, Central Otago 158
Kohatu Tunnel 92, 193
Komata Gold Mines 113
Konka Board 130, 176
Kopuawhara Valley *226*
Kopua Viaduct, Hawkes Bay 94, *97*
Kotuku, West Coast 168, *169*
Kowhai River, Kaikoura 39
Kreijmborg, Father Charles 193, 194, *195*
Kurow, Waitaki Valley 203, *209*

Lancaster (Vickerman & Lancaster) 216, *219*
Laing-Meason, Gilbert 132, *132*
Lake Coleridge station, Mid-Canterbury 23, *23*
Lake Coleridge Hydro-electric Station 134, 154, *155*, *164*, 189
Lake Brunner 215
Lake Dunstan 156
Lake Ratapiko 187
Lake Taupo 193
Lake Waikare 55
Lake Waikaremoana 204
Lake Wakatipu 190
Lambot 17
Langevard, H. 98, 99
Langley Dale station, Marlborough 9, 28
Lawrence 32, *34*, 38
Leaning Rock Creek Bridge, Central Otago 156, *166*
Le Corbière Lighthouse, Channel Islands 71
Le Corbusier 18
Lendon run, Mid-Canterbury 54, *59*
Levels Road Board 53
Levels station, South Canterbury 29
Lewis, C.A. 102
Lime concrete, 22, 31, *31*, *81*, 86
Lime, hydraulic 80, 87
Logan Bank, Auckland 26
Limestone Island, Whangarei Harbour 89
Luck, Isaac, architect 42
Luttrell, Alfred 114, 115, *115*
Luttrell brothers 114, 170, *171*, 180
Luttrell, Sidney 114, 170, *181*
Lyall Bay, Wellington 168, 216, *216*, 217
Lysnar, William D. 148

MacArthur-Forrest 90
Macdonald, C.Fleming 21, 102, 115, *115*
Macdonald, James 88
Macdonald, John A. 178
Macgeorge, Leslie D. 69
McLean, Donald 29
McLean, Hugh 76, 77, *85*
McLean, Hugh H. 77
McCurdie, W.D.K. 174
McFadden, W.and Son 215
McGill, David 42
McGowan, Hon. James 98
McGregor, John. 56
McMillan, Dr D.G. 204

Magazine, Lyttelton Harbour 10
Mahoney, Edward and Sons 130
Mahoney, Father 171
Mahurangi River 86, 88
Mahurangi limestone 87
Maillart, Robert 19, 187
Main Highways Board 210
Mair, John T., Government Architect 214, *215*
Makaraeo, North Otago 88
Makara *133*, 168, *171*
Makara Road Board 134
Makarau River 223, *223*
Maketawa Dairy Factory 105
Manakau, Manawatu 159, *159*
Manawatu Catchment Board 159
Manawatu River 94, 158, *166*, 223
Mangahao Hydro-electric Station 189
Manganui River, Taranaki 104, *108*, 128, *130*, *186*, 187
Manganui-o-te-Ao Valley 228
Mangaotuku Stream, Taranaki 54
Mangapurua Bridge (Bridge to Nowhere) 223, *223*, 228
Mangatainoka 212, *212*
Mangatainoka River 212, *220*, 222
Mangawhai, Northland 76
Mangere, Auckland 69, 70, *70*, 148
Manning, H.E. 171
Manorburn Dam, Central Otago 158, *166*
Manuherikia River, Central Otago 225, *229*
Manukau City 177, 222, *224*
Maori 222
Maraekakaho station, Hawkes Bay *100*, 101
Marchbanks, James 144
Marine Department 71, 72
Marlborough 9, 28
Martha Mine, Waihi *106*, 112, 113, 122
Martinborough 132, *132*
Marton 73, *83*, 117
Mason and Wales 39, 62, 65, *65*
Matamata 26, 65, *66*
Matangi Dairy Factory, Waikato 156
Maungaroa station, Bay of Plenty 135
Mauriceville County Council 172, 184
Maze House, Pleasant Point 53, *53*
Medway Valley, England 14
Menai Straits Bridge, Wales 16
Mendelsohn, Erich 18
Mercer, Archibald 40, 41
Mercer Road Farmhouse, South Otago 40, *41*
Messenger, Frank 188, *188*
Messenger, Griffiths and Taylor 188
Meyer, Charles 42
Mid-Canterbury 23, *23*, 50, *50*, *51*, *52*, 53
Midland Railway 122, *127*
Milburn Lime and Cement Company Ltd 88
Miller, W.A. 172
Miller's Building, (Civic Offices) Christchurch *228*, 229
Mill Hill Fathers 193
Ministry of Works and Development 74, 203, 222
Miramar Peninsula, Wellington 74
Mitchell, John 130, *131*
Model Building By-Laws 213
Mokohinau Lighthouse 71, 123, 147
Monier, Joseph 17
Moodie, Wilson & 118
Moore, George, H. *46*, 49, 62, *63*, 64, *64*
Moore, R.F. 105, 120, *125*, *126*

Moorhouse, Sefton 115
Morris, William 53
Morton, W.H. 115, *116*, 128
Morton Dam, Wainuiomata 128
Mosgiel 23, 25
Motukawa Powerhouse, Taranaki 187
Motupipi, NW Nelson 88
Mouchel, L.G. 18
Mount Eden, Auckland 128
Mountfort, Benjamin W. 29, 30, 42, 43, *43*
Mountfort, Cyril.J. 144
Mulvihill, John 70
Murchison Earthquake 210

Naenae, Hutt Valley 94
Napier 130, 144, 159, 192, *196*, 212, 213, *213*
Napier Brewery Coy Ltd 214
Napier Central Fire Station *213*, 214
Napier Town Planning Committee 213
Napier Earthquake (Hawkes Bay Earthquake) 159, 162, 196, 213
National Roads Board 210
Natusch, Charles T. 115
Nelson 73, 193
Nelson College 210, *212*
Nelson-Glenhope Railway 193
Nelson, Moate and Coy Building, Wellington 115
Nervi. Pier L. 19
Newnham, William L. 220, *221*
New Plymouth 54, *54*, 55, 102, 123, *188*, 188
New Zealand 15, 22, 23, 26, 66, 84, 90, 99, 101, 128, 154, 156, 158, 159, 172, 174, 188, 193, 202, 204
New Zealand and Australian Land Company 24, 29, 56, 58
New Zealand Cooperative Dairy Company 156
New Zealand Express Company Building, Dunedin 115, *115*
New Zealand Express Company 21, 114, *115*, 115
New Zealand Historic Places Trust 24, 42, 69, 94, 98, 120
New Zealand Loan and Mercantile Agency Company Limited 42
New Zealand Portland Cement Company 89, 117
New Zealand Railways 126
New Zealand Railways Workshops 66, *67*
Ngaere, Taranaki *146*, 147
Ngaire Cooperative Dairy Company 146
Ngamuwahine Bridge, Bay of Plenty 210, 222
Ngaruawahia 55, *57*, 66
Ngatea 177
Ngatoroiti Stream Bridge, Taranaki *103*, 105
North Island 130, 202, 204
North Canterbury 49, 56, *57*, *96*
North East Valley, Dunedin 91, *91*
Northern Roller Milling Coy, Auckland 112
Northfleet, England 13
Northland (North Auckland) 76, *77*
Northland Tunnel, Wellington 193
North Otago *11*, 39
North West Nelson 88
Nôtre Dame du Raincy, France 19

Oamaru 30, *30*, 91, 92, *92*
Oamaru Breakwater *30*, 30, 31

Oamaru Harbour Board 31
Ocean View Hotel, Dunedin 38
Ohakea Air Force Station, Hangars 230, *231*
Ohinemuri Goldfield 134
Ohinemuri River 93
Okarito, West Coast 24
Okuku Pass station, North Canterbury 56, *57*
Old Mountain Road, Taranaki 104
Omakau Bridge, Central Otago 225, 228, 229
Onehunga 148
Opawa River Bridge, Blenheim 158, *167*, 220, 222
Ophir, Central Otago 224
Opiki, Manawatu *157*, 159
Opunake 68, 69
Ormondville 94, *97*
Orongorongo Range 128
Otago 24, 25
Otago Harbour 75
Otago Harbour Board 114
Otautau, Southland 98
Otautau Roller Flour Mills 98
Outram 24, 28, 62
Overton, Rangitikei 73, *83*, 117
Oxford, Canterbury 53

Pachuca Tanks 119
Pahiatua 158, *220*, 222
Pahiatua County Council 222
Palmerston North 155, 223
Pantheon, Rome 12
Pareora Estate, South Canterbury 56, 58
Parker, Rev. James 13
Parker, Obadiah 14
Parkhurst, I.O.W., England 14
Patea 222, *222*
Patea River 147, 188
Paterson, A.J. 188, *189*
Patrick Street, Petone *100*, 101
Paul J. and Coy 54, *54*
Pawarenga, Northland 193
Peel Street Bridge, Gisborne 178
Pelichet Bay, Dunedin 88
Perret, Auguste 19
Petone *100*, 101
Petre, Francis W. 43, *44*, *45*, *47*, 48, 49, 88, 91, *91*, 92, *92*, 101, *111*
Pettigrew House, Opunake 68, 69
Phillips, Eric 192, *196*
Piakau Creamery, Taranaki 105. *105*
Pipi Bank station, Hawkes Bay 92
Pipiriki 223
Piriaka 190
Plaster of Paris 15
Pleasant Point 53, *53*
Plunket Street Wall, Wellington *204*, 204, 205
Point Gordon, Wellington 74, 75, *75*
Pond, J.A. 87
Pont de Nôtre Dame, France 13
Port Chalmers 31, *113*, 114
Porter and Martin 128, *129*
Portland, Northland *175*, 176
Portland cement (see Cement, Portland)
Post and Telegraph Department 122
Potter, Ernest H. 145
Poverty Bay Farmers' Meat Company 148, *151*
Princes Wharf and Sheds, Auckland *180*, 187
Progress (periodical) 101, 118, 122
Property and Finance Company, Invercargill 123

Public Toilet and Tram Shelter, Wellington *205*, 206
Public Trust Office, Napier 192, *196*
Public Works Department 20, 21, 30, 71, 75, 157, *157*, *158*, 159, *160*, 160, *166*, *167*, 172, *173*, 188, 190, 203, *221*, 223, *223*, *226*, 228, *229*, 230, 231, *231*
Puhinui Stream, Wiri, Manukau City 210
Puhoi *175*, 176
Pumice concrete 110, 114
Pumphouse No 5, Martha Mine, Waihi *106*, 112, 113
Putaruru 202
Putoli, Italy 12

Queenstown 160, 161

Radnor Road, Taranaki 104
Raes Junction, Central Otago 172
Raetihi 228
Railways Department (New Zealand Railways) 66, 122, 126
Railway Wharf, Auckland 112
Rakaia Bridge 230
Rangiora 42, *43*, 43, 58, 130, *143*
Ransome, Ernest L. 18
Ransome, Frederick 18
Rawhitiroa Road Bridge, Taranaki 147, *147*
Reinforced concrete 10, 16-20, 98 et seq
Renaissance 19, 48, 101, 110, 202
Retaining Wall, Upland Road, Wellington 217, *217*
Rhenish trass 13
Richardson and Company 170
Richardson, Greer and Company 161
Riccarton Racecourse Stand, Christchurch 170, *171*
Rimutaka Railway 43
Ripapa Island, Lyttelton Harbour 75
Risorgimento Bridge, Rome 18
Riwaka, North West Nelson 73
Robinson, E.C. *103*, 104, 105, *108*, 130
Roche lime 86
Rock-faced concrete blocks 174
Roman Empire 13
Romanesque architecture 49, 72, 168
Ross, Alex 75
Ross, David 32, *34*, 38
Rotokakahi, Northland 193, *195*
Rotorua 110, 113, 114, 214, *215*
Rotorua City Council 114
Royal Institute of British Architects 10
Royal New Zealand Navy 77, 78
Ruakiwi Road Reservoir, Hamilton 215, *218*
Ruamahanga River, Wairarapa 132, *132*
Russia 15
Russian war scare 74
Rutherford, Baron 210
Rutherford House, Nelson College, Nelson 210. 211, *212*
Rutherfurd and Coy 89
Ryburn's Bridge, Clevedon 210, *224*

St Andrews, South Canterbury 56, 58,
St Clair, Dunedin 47
SS Cletus and Remigius, Waihou, Northland 194, *195*
St Dominic's Priory, Dunedin *44*, 49
St James' Church, Auckland 32, *32*, 35,
St Jean de Montmartre, Paris 17
St John the Baptist Church, Rangiora 42, *43*

St John's Church, Barrhill, Mid-Canterbury 54, *60*
St Joseph's Cathedral, Dunedin 49
St Jude's Church, Wellington 168
St Mary's Church, Southbrook, Rangiora 58, *60*
St Mary of the Angels Church, Wellington 171, 172, *183*
St Mary's Church, Karori, Wellington 128, *129*
St Matthew's Church, Hastings *139*, 144
St Patrick's Basilica, Oamaru 91, *92*
St Paul's Hall, Kaikoura 90, *91*
St Peter's Church (Cathedral) Hamilton 154, 155
Sacred Heart Church, North East Valley, Dunedin 91, *91*
Sandford and Brown 228
Saunders, William 98
Saylor, David O. 16
Scott, George G. 24
Schirmer Factory, New York 112
School Dental Clinic, Wellington *229*, 230
Scratchley, Lt Col P.H. 74
Seaman and Hurry 88
Sea Wall, Lyall Bay, Wellington *216*, 216
Sea Wall, St Clair, Dunedin 135, *142*
Seaton, Sladden and Pavitt 220, 222
Shand, James 28, 62
Shotover River, Central Otago 98, 159
Siacci 233
Silver, S.T. 140, 146
Sinclair-O'Connor, A. *138*, 145, *161*, 162
Sir W.G.Armstrong, Whitworth & Co Ltd 202
Skippers Bridge, Central Otago 98, *109*
Smeaton, John. 13
Snell, Florence 86
South Africa 15
South America 15
Southbrook, Rangiora 58, *60*
South Canterbury 8, 24, 42
South Otago 41, 64, 65
Southgate, John 86
Southland 62, 94, *95*, 98, 124
Southland Girls' High School 124
South Island 39, 114, 203
South Taranaki 146
Spanish Mission architecture 129, 211
Speechley, Robert 24
Stadium Communale, Florence, Italy 19
Stanley, Christopher 12
Stanton, William W. 145
Starr and Company, San Francisco 18
Staveley, N.Crofton 194, *199*
Stevenson, Robert L. 157
Stevenson, D.& T. 157
Stewart, Dr McBean 50, *50*
Stillwater, West Coast 215
Stewart, Isabella 172
Stewart, Robert 172
Stewart Creek Bridge, Island Block 172, *184*
Stussi and Whitney 20
Stratford Borough 186, 188
Strathconan, near Fairlie 29, *29*, 30
Sunnyside Mental Hospital, Christchurch 29, *30*
Swan, John S. 115
Swansea, Wales 18
Swyncombe station, Inland Kaikoura 40, *40*
Sydney, Australia 22
Synagogue, Auckland 72, *81*

T & G Building, Wellington 194, *201*, 202
Taieri 92, 93
Taihape 148, 150
Tall, Joseph 17
Tane Hemp Company Ltd *157*, 159
Tangahoe River 228
Tangahoe Valley Bridge, *226*, 228
Tara, Northland 76, *77*
Taraheru River, Gisborne 178
Tarakohe, N.W. Nelson 88
Taranaki 54, *54*, 102, *103*, 105, *105*, *108*, 130, 146, 147, *147*, *186*, 188, *188*
Taranaki Brewery, New Plymouth 54, *54*
Taranaki County Council 102, 104, *108*, 128, 130
Taranaki Electric Power Board 186
Tariki Road, Taranaki 104, *108*, 186, *186*, 187
Tarureka, Wairarapa 24
Tata Islands, Golden Bay 88
Taumarunui Borough Council 190
Tauranga County Council 210
Tauroa, Havelock North 152, 153, *153*, 154, *192*, *193*
Taylor, John E. 69, *70*, 70
Taylor, J.P. 62
Te Aroha 190, *191*, 210
Te Awa, Waikato 75
Te Kaha, Bay of Plenty 135, 136, *137*
Te Kauwhata 90
Te Kuiti 89
Telford, Thomas 16
Temuka 73, *82*
Thacker, Dr T. 178
Thames 174
Thames School of Mines 76
Thames Estuary, England 13
The Brothers run, North Canterbury 58
'The Cliffs', Dunedin *47*, 48
The Elbow run, Southland 62
The Scientific New Zealander (Progress) 101, 118, 122
The Elms run, Inland Kaikoura 36, 90
'The Taj', Wellington *205*, 206
The Temperance and General Mutual Life Society 194, *201*
Thompson House, Invercargill 124
Thorndon, Wellington 31, *31*
Timaru 147
Timber kiln 168, *169*
Tokomaru Bay, East Coast 148
Tolaga Bay Wharf 202, *207*
Tolaga Bay Harbour Board 202, 206
Tongue Point, South Karori 147
Totara Estate, North Otago 24
Transit New Zealand 210
Trentham 170, *181*
Trigg, W.J. 113
Triumph Dairy Factory, Ngaere 146, *146*, 147
Troup, George A. 94, *97*
Tuai, Lake Waikaremoana 204
Tuakau 222
Tuakau Bridge *220*, 222
Tuapeka County Council 184
Tuapeka West *185*, 186
Tudor Towers, Rotorua *110*, 114, 214
Tudor Renaissance architecture 48
Tui Brewery, Mangatainoka 212, *212*
Turanganui River, Gisborne 178, *179*
Turner,C.A.P. 19
Turner, Charles W.O. 230, *231*

United Kingdom 149
United States of America 16, 123, 154, 174

University of Otago 23
Upland Road Wall, Wellington 217, *217*
Upper Dowling Street Steps, Dunedin *191*, 192
Upper Hutt 180
Upper Moutere 74, *93*
Upper Shotover River, Central Otago 108, *109*
Upper Skagit Valley, USA 176

Valve Tower, Karori Reservoir, Wellington 31, *33*
Van Asch, Ivan 162
Van Nelle Factory, Holland 19
Venus Street houses, Invercargill *121*, 124
Vickerman, Neville L *180*, 187
Vickerman, Hugh (Vickerman and Lancaster) 216, *219*
Victoria Battery, Waikino 96, 134
Vitruvius 12

Wadestown, Wellington *41*, 42
Waiau River 94, *95*
Waiau Stream Viaduct, Kopuawhara Valley *226*, 231
Waihenga Bridge, Wairarapa 132, *132*
Waihi *106*, *119*
Waihi Gold Mining Coy *106*, 112, 113, 120, 134
Waihou, Northland 194, *195*
Waikato 75, 90
Waikato River 202, *208*, *220*, 222, *227*, 229
Waikino 93, *96*, *112*, 120, *133*
Waimakariri Gorge Bridge, Canterbury 51, *52*, 53
Wainuiomata 128
Waiongongati Stream, Taranaki *103*, 104
Waipa County Council 156
Waipa River, Waikato 55
Waipaoa, Poverty Bay 148, 149, *151*
Waipara, North Canterbury 49
Waipuku Stream, Taranaki 104
Wairarapa 24, 132, *132*, 135, 172, *184*, 212, 212, 230
Wairio Stream, Otautau, Southland 98
Waitaki Dam 204, *209*
Waitaki Hydro Medical Association 204
Waitaki River, North Otago 204, 208
Waitara River, Taranaki 187
Waitepeka, South Otago 40, *41*
Waiwakaiho River, New Plymouth 22, 102
Wakefield, England 14
Waldie, A.B. 188
Wales, N.Y.A. (Mason and Wales) 39
Wallace County Council 94
Wanganui 54, 194, *199*
Wanganui Racing Club *11*, 42
Wanganui River 190
Ward, Sir Joseph 94
Warkworth 75, 86, 89, *99*, *107*, 116, 223
Warnock, Richard 76, *84*
Warnock, May 76, *84*
Warren, John 154, 155
Wason, John C. 53, *59*, *60*
Water Towers 66, *67*, 69, *141*, 146, 194, *199*, 215, *218*
Weaver's Granary and Flour Mill, Swansea, Wales 18
Wellington 31, *31*, *33*, 41, 42, 55, 74, 75, 94, 115, *116*, *117*, 118, 128, *129*, *131*, *133*, 134, 144, *163*, 171, *171*, 172, *183*, 188, *189*, 194, *201*, 202, *204*, *205*, 206, *216*, 216, 217, *217*, *229*, 230

Wellington City Council 128, 188, 206
Wellington Racing Club 170, *181*
Wellington Waterworks Act 31
Wellington Philosophical Society 144
Werner, P.N. 203
Wesley Chambers, Hamilton 189
West Coast 24, 92, 168, *169*, 170
Western Springs, Auckland 77, 130
Westshore Bridge, Napier 159
Whangape Harbour, Northland 193
Whangarei Harbour 89
Whare Ra, Havelock North 152, *152*
Wheatport, USA 18
Whenuapai Air Force Station 21, 230, *231*
White, John B. 14, 15
White, John Bazely and Sons 14, 15
White, J.B. and Brothers 15
White, Leedham 15
White, Thomas H. 55, *57*, 66
Whiterock station, North Canterbury 56, 90, *96*
Whitney, Stussi and 20
Wilkinson, William B. 16
Williams, Cyrus J.R. 202, 207
Williams, E.A. *213*, 214
Wilson, James 86
Wilson, John 86
Wilson, John and Coy Ltd 86, 88, *99*, *107*
Wilson, Nathaniel 86, 87, 93, 100
Wilson, W.J. 87
Wilson's (NZ) Portland Cement Coy 89, *175*, 176
Wilson's Portland Cement Cement Coy 89, 174,
Windbreak, Makara Hill, Wellington *133*, 134
Windermere, Mid-Canterbury 50, *51*
Winstone, George 89
Wireless Road, Kaitaia 136, *136*
Wohlmann, Dr Arthur S. 113
Woodside, Dunedin 43, *45*, 48
Workers' Dwellings Act *100*, 101
Worley, Rupert 215, 219
Worthington 53
Wright, Edward G. 50, *51*, 53
Wright, Dr Frederick 70
Wrigley, J.W. 113

Young, W.Gray, 211, *212*
Yugoslavia 12